FEMINIST DIGITAL CITIZENSHIP IN AFRICA

Digital Africa series, edited by Tony Roberts

Digital Africa explores how digital technologies have opened new spaces for the exercise of democratic rights and freedoms in Africa, as well as how repressive regimes have used digital technologies to monitor, diminish or remove those rights. The series foregrounds vital new research from across the continent that combines empirical rigour with theoretical sophistication in order to offer new, more nuanced perspectives on the interactions between digital technology and social life on the continent. In so doing, it offers an important, in-depth corrective to existing studies of the relations between digital technologies and social and political power, studies that have overwhelmingly focused on the Global North.

Each volume is co-edited by the series editor in collaboration with an established scholar in the field, and each features comparative and rich case studies authored by ECRs and other African scholars involved in the African Digital Rights Network. Thanks to contributors' diverse disciplinary and professional backgrounds, each chapter offers uniquely practical suggestions for policy, legislation and practice.

Titles include:
Digital Citizenship in Africa: Technologies of Agency and Repression
Digital Disinformation in Africa: Hashtag Politics, Power and Propaganda
Digital Surveillance in Africa: Power, Agency, and Rights
Internet Shutdowns in Africa: Technology, Rights and Power
Feminist Digital Citizenship in Africa: Agency, Rights and Resistance

FEMINIST DIGITAL CITIZENSHIP IN AFRICA

Agency, Rights and Resistance

Edited by
Tanja Bosch and Tony Roberts

ZED
LONDON • NEW YORK • OXFORD • NEW DELHI • SYDNEY

ZED BOOKS

Bloomsbury Publishing Plc, 50 Bedford Square, London, WC1B 3DP, UK
Bloomsbury Publishing Inc, 1359 Broadway, New York, NY 10018, USA
Bloomsbury Publishing Ireland, 29 Earlsfort Terrace, Dublin 2, D02 AY28, Ireland

BLOOMSBURY and Zed Books are trademarks of Bloomsbury Publishing Plc

First published in Great Britain 2025

Copyright © Tanja Bosch and Tony Roberts, 2025

Tanja Bosch and Tony Roberts have asserted their right under the Copyright, Designs and Patents Act, 1988, to be identified as Authors of this work.

For legal purposes the Acknowledgements on p. xviii constitute an extension of this copyright page.

Series design by Toby Way

This work is published open access subject to a Creative Commons Attribution-NonCommercial-NoDerivatives 4.0 International licence (CC BY-NC-ND 4.0, https://creativecommons.org/licenses/by-nc-nd/4.0/). You may re-use, distribute, and reproduce this work in any medium for non-commercial purposes, provided you give attribution to the copyright holder and the publisher and provide a link to the Creative Commons licence.

Bloomsbury Publishing Plc does not have any control over, or responsibility for, any third-party websites referred to or in this book. All internet addresses given in this book were correct at the time of going to press. The author and publisher regret any inconvenience caused if addresses have changed or sites have ceased to exist, but can accept no responsibility for any such changes.

A catalogue record for this book is available from the British Library.

Library of Congress Cataloging-in-Publication Data
Names: Bosch, Tanja Estella, editor, author. https://id.oclc.org/worldcat/entity/E39PCjvX7DGkWJkXmJhW7c4C33 | Roberts, Tony, 1963- editor, author.
Title: Feminist digital citizenship in Africa : agency, rights and resistance / edited by: Tanja Bosch and Tony Roberts.
Other titles: Digital Africa (Series)
Description: New York : Zed Books, 2025. | Series: Digital Africa | Includes bibliographical references and index. | Summary: "This open access edited collection offers the first-ever book-length volume on feminist digital citizenship in Africa. It offers multiple, theoretically grounded case studies by African researchers covering countries across the length and breadth of the continent, including non-majority-English countries"– Provided by publisher.
Identifiers: LCCN 2025021927 (print) | LCCN 2025021928 (ebook) | ISBN 9781350500525 (hardback) | ISBN 9781350500488 (paperback) | ISBN 9781350500518 (epub) | ISBN 9781350500501 (pdf)
Subjects: LCSH: Digital communications–Political aspects–Africa. | Social media–Political aspects–Africa. | Women–Political activity–Africa. | Political participation–Africa. | Feminism–Africa. | Citizenship–Africa.
Classification: LCC HM851 .F457 2025 (print) | LCC HM851 (ebook) | DDC 302.231096–dc23/eng/20250507
LC record available at https://lccn.loc.gov/2025021927
LC ebook record available at https://lccn.loc.gov/2025021928

ISBN: HB: 978-1-3505-0052-5
PB: 978-1-3505-0048-8
ePDF: 978-1-3505-0050-1
eBook: 978-1-3505-0051-8

Series: Digital Africa

Typeset by Integra Software Services Pvt. Ltd.
Printed and bound in Great Britain

For product safety related questions contact productsafety@bloomsbury.com.

To find out more about our authors and books visit www.bloomsbury.com and sign up for our newsletters.

BLOOMSBURY OPEN ACCESS

An ebook edition of this book is available open access on bloomsburycollections.com. Open access was funded by the Bloomsbury Open Collections Library Collective. Bloomsbury Open Collections is a collective-action approach to funding open access books that allows select authors to publish their books open access at no cost to them. Through this model, we make open access publication available to a wider range of authors by spreading the cost across multiple organizations while providing additional benefits to participating libraries. The aim is to engage a more diverse set of authors and bring their work to a wider global audience. More details, including how to participate and a list of contributing libraries, are available from http://www.bloomsbury.com/bloomsbury-open-collections.

*To Evelyn Rosie Bosch, who made it impossible for me to be anything but a feminist.
To Chianah and Etta, for filling me with pride and joy.*

CONTENTS

List of Illustrations   xi
List of Contributors   xii
Acknowledgements   xviii

Chapter 1
THEORIZING FEMINIST DIGITAL CITIZENSHIP IN AFRICA   1
   Tanja Bosch and Tony Roberts

Chapter 2
QUEER FEMINIST DIGITAL CITIZENSHIP ON NIGERIA'S X: THE CASE OF #UJUANYA   23
   Ochega Ataguba

Chapter 3
'OF COURSE WE ARE ANGRY': LUSAKA WOMEN AND THE ZAMBIAN FEMINISTS FACEBOOK PAGE   51
   Chishimba Kasanga and Priscilla Boshoff

Chapter 4
KEEPING EACH OTHER SAFE: TRANSGENDER PEOPLE'S ONLINE SOLIDARITY STRATEGIES   69
   Nyx McLean

Chapter 5
DIGITAL FRUGALITY IN LOW-INCOME COMMUNITIES IN SOUTH AFRICA: ENABLING WOMEN'S CITIZENSHIP OF SURVIVAL   91
   Alette Schoon and Marion Walton

Chapter 6
#GUINEENNEDU21ESIECLE AND THE RADICAL POTENTIAL OF FEMINIST ACTIVISM IN CONTEMPORARY GUINEA   109
   Clovis Bergère

Chapter 7
TRANSFORMATIVE MOMENTS IN FEMINIST DIGITAL CITIZENSHIP IN POST-REVOLUTION EGYPT   127
   Manal Hassan

Chapter 8
DISMANTLING BOUNDARIES: MOZAMBIQUE'S TRAILBLAZING
FEMINIST DIGITAL CITIZENSHIP   151
  Lissungu Mazula and Dércio Tsandzana

Chapter 9
DIGITAL FEMINIST CITIZENSHIP IN MALAWI: MWIZA CHAVURA'S
RAPE SONG   169
  Jones Maweranga and Godwins Lwinga

Chapter 10
CONTESTING BOUNDARIES: GENDERED CITIZENSHIP AND DIGITAL
ACTIVISM IN SUDAN   183
  Maha Bashri

Chapter 11
DIGITAL FEMINISM IN ETHIOPIA   201
  Selamawit Tezera Chaka

Index   217

ILLUSTRATIONS

2.1	A visual breakdown of data screening and selection shows the tweets in the dataset for #UjuAnya	35
2.2	Threatening response to @UjuAnya's tweet	37
2.3	Homophobic comments on @UjuAnya's tweets	42
2.4	Homophobic tweets on X	43
7.1	Timeline of the Egyptian feminist movement in relation to the timeline of the global feminist movement	133
8.1	Internet use in Mozambique (free to use, DataReportal 2025)	159

CONTRIBUTORS

Ochega Ataguba is a PhD student at the Centre for Film and Media Studies, University of Cape Town, South Africa. She previously completed a master's degree in mobile communication research. Her thesis, 'Power, Asymmetries and Support Networks: Zimbabwean female migrant domestic workers and ICTs', draws on intersectional lenses to intervene in the linear narratives of female domestic workers as perpetual victims to demonstrate their new practices of subversion and autonomy through mobile phone use. Her wider areas of interests include public health messaging in the digital era, citizenship and politics, gender and politics of identity, ethnicity, culture and African feminist dissent in social media spaces.

Maha Bashri is Associate Professor of Communication at the Africa Institute in Sharjah, United Arab Emirates, specializing in intersectional research on gender, social justice and marginalized voices. Born in Sudan and an immigrant to the United States, her unique positionality informs her work on the impact of media and communication on social change across various geopolitical contexts. Dr Bashri's research has been published in prestigious journals, such as *Information, Communication & Society*, where her article 'Networked Movements and the Circle of Trust: Civil Society Groups as Agents of Change in Sudan' (2021) examines the role of digital media in grassroots mobilization. Her co-edited book, *Minority Women and Western Media: Challenging Representation and Articulating New Voices* (2020), highlights her commitment to amplifying marginalized perspectives. Dr Bashri actively pursues equity and diversity in the field, challenging prevailing narratives and promoting diverse perspectives on a global scale.

Clovis Bergère is Assistant Director for Research at the Institute for Advanced Study in the Global South, Northwestern University in Qatar. He previously worked at the Center for Advanced Research in Global Communication (University of Pennsylvania), where he also completed a postdoctoral fellowship. His research examines the politics of youth in Guinea, West Africa, focusing on digital media and urban life. He holds a PhD in Childhood Studies from Rutgers University (2017) and has taught at Rutgers University, the University of the Arts and the University of Pennsylvania. His work has appeared in the *International Journal of Communication*, *Public Culture*, *African Studies Review* and *Journal of Childhood Studies*, as well as several edited volumes. Prior to returning to academia in 2011, he worked as a project manager in Children's Services in London, where he oversaw the building of over thirty playgrounds and youth centres focused on natural play and collaborative design.

Tanja Bosch is Professor of Media Studies and Production in the Centre for Film and Media Studies at the University of Cape Town, South Africa, where she is also the National Research Foundation Research Chair in Digital Media Sociology. Dr Bosch has published widely in the field of radio studies in South Africa and is emerging as a leading African academic publishing in the area of social media activism, with her work on #RhodesMustFall and #FeesMustFall and other hashtagged campaigns. She is the author of *Broadcasting Democracy: Radio and Identity in South Africa* and *Social Media and Everyday Life in South Africa*, and co-editor of *Digital Citizenship in Africa*. Tanja is the Chairperson of the African Digital Rights Network.

Priscilla Boshoff is Associate Professor and Deputy Head of School at the School of Journalism and Media Studies, Rhodes University. She teaches media and cultural studies at undergraduate and postgraduate levels, with an emphasis on the intersection of gender and popular culture in Africa. Much of her research, which draws on constructivist and decolonial frames, has focused on the representations of women and marginalized communities in tabloid newspapers, television and social media. Recent publications include 'Murders of gay men in South Africa: an analysis of #JusticeFor … hashtags', an examination of the construction of the murders of LGBTQ persons on Twitter, and 'New mythologies: violence and colonialism in Hollywood blockbusters', a critique of Hollywood's representations of decoloniality.

Selamawit Tezera Chaka is a feminist researcher with a robust background in development economics and extensive experience working at the intersections of youth, gender, human rights and technology for over six years at national, regional and international organizations in different capacities. Her passion is making online space accessible and safe for women and empowering them to use it. She also supports regional feminist organizations and collectives as a global advisor for FRIDA| The Young Feminist Fund. As a Safe Sisters fellow, she provided digital security training to over one hundred women human rights defenders, women journalist and young feminists, and has recently founded sheEsecures, a digital platform that advocates for women's digital rights and internet freedom. Selamawit is a Mandela Washington Fellow.

Nancy Gakahu is a lecturer of Journalism and Mass Communication at Masinde Muliro University in Kenya. She holds a PhD in Digital Politics within the context of Kenya from the University of Leeds, UK. Nancy's research interests include gender and communication, digital politics, communication and democracy, political communication, and citizenship in transitional democracies. She is a current reviewer of the *Journal of Information, Communication and Society*. Nancy is also an active member of the International Association of Media and Communication Research as well as the Academic Association of the United Nations System.

Manal Hassan is a gender and technology expert with a robust background in IT and extensive experience working with various NGOs and initiatives. Her passion is making technology accessible to non-techies and empowering them to use it independently. She has a wealth of experience in facilitation, community building, and working on web technologies and new media with NGOs, human rights activists and children. Manal co-founded the Egyptian GNU/Linux Users Group in 2004, the Arab Techies Collective in 2008 and Open Egypt and Motoon in 2013, serving as Motoon's Executive Director for seven years. Throughout her work in building Egyptian and Arab tech communities, she has been dedicated to enhancing women's participation in technology, striving to create safe and welcoming environments, communities and spaces for women. As the mother of a neurodivergent son, autism and special needs education became part of her everyday life, thus making diversity and inclusion central to all her professional endeavours.

Chishimba Kasanga is a Rhodes Scholar reading for a Doctor of Philosophy (DPhil) in Geography and the Environment at the University of Oxford. Her DPhil research seeks to understand how the biographies of Alice Lenshina, Julia Chikamoneka and Sara Longwe can provide insights into how Zambian women have historically used their bodies as tools in resistance to patriarchy. Before Oxford, Chishimba was a Beit Scholar and pursued an MA in Journalism and Media Studies at Rhodes University in South Africa. The book chapter was inspired by her master's research in which she wanted to understand how Zambian women fans of the Zambian Feminists Facebook page negotiate their strong online feminist identity and their offline identity as mothers, sisters and daughters living in a patriarchal society. Her research interests include African Feminism(s), gender, identity, the body and digital media activism.

Godwins Lwinga has lectured in Media Studies at the Malawi Institute of Journalism (MIJ), Islamic Philosophy at Domasi College of Education, and Ethics, Philosophy of Religion and African Philosophy at Mzuzu University and University of Livingstonia. He received his BA from African Bible College, his MA from Mzuzu University and his PhD in Leadership Ethics from Evangelische Theologische Faculteit in Leuven, Belgium, in 2023. His research focused on the ethics of uMunthu and of Dietrich Bonhoeffer as resources for a new responsibility-based anti-corruption approach in Malawi. He has a passion for interdisciplinary research, particularly about religion and environmental ethics, organizational leadership and ethics, anti-corruption studies, Malawi's one-party state political ethics, digital feminism and media ethics. He has published in the *Evangelical Review of Theology* and *MAGU Ethene Journal*. He is finalizing editing his book, *Christianity and the Ndali*, to be published by Theological Society of Malawi Press.

Jones Hamburu Mawerenga is a Senior Lecturer in Systematic Theology and Christian Ethics at the University of Malawi. He is also a Research Fellow in the

Department of Practical Theology and Mission Studies, Faculty of Theology and Religion, University of Pretoria, RSA. His research interests and publications include religion and sexuality, albinism, theological education in Africa, the Covid-19 pandemic in Africa and African feminist epistemology. The current chapter was inspired by the quest to understand how Malawian women exercise their digital citizenship by resisting a salient rape culture that is tolerated in society. In this endeavour, Malawian women have demonstrated resilience, self-mobilization and agency in fighting against the rape culture and everything that jeopardizes their dignity and well-being.

Nyx McLean is a rated transdisciplinary scholar in the fields of LGBTIAQ+ identities, digital technology and social movements. Their scholarship seeks to trouble the intersections of technology and power, with the intention of envisioning radically inclusive realities through a queer feminist intersectional lens. Notable research contributions include an exploration of transgender, non-binary and gender-diverse peoples' experiences of online gender-based violence, and a study of community-led participatory AI, in the case of Queer AI. Additional research focus areas include Southern technologies, digital safety and security for marginalized communities, collaborative and participatory AI methodologies, digital counter publics, online communities and human–digital interaction. Nyx McLean is the National Research Chair at Eduvos and a Research Associate at Rhodes University in the School of Journalism and Media Studies.

Lissungu Sheila Mazula holds a master's degree in Gender and Development Studies from the University Eduardo Mondlane, where she also earned her bachelor's degree in Political Science. Working with local women activists, in her master's dissertation, Lissungu explored the emergence of feminist movements on the internet. She also used art as a form of feminist expression, through painting and the defence of women's blackness. With extensive experience collaborating with United Nations organizations (UNICEF, World Food Programme, and WHO), she has successfully managed numerous programmes focused on promoting gender equality and women's empowerment. Her research primarily focuses on feminist citizenship, intersectionality and female engagement.

Chris Paterson is Professor of Global Communication at the University of Leeds. Since 2019, he has led research with colleagues in Kenya and Ghana about climate change adaptation by women in rural areas and the ways climate change is communicated in a fast-changing information environment. He has authored books about news agencies and war reporting, has published eight anthologies and over thirty articles and chapters, and has led the International Communication Association's Global Communication and Social Change division. Current research focuses on climate communication, digital colonialism and neo-imperialism, particularly in an African context. At the University of Leeds, he teaches international communication and climate communication, and he also teaches for the Leeds Centre for African Studies.

Tony Roberts is a Research Fellow at the Institute for Development Studies (IDS), UK. After a period as lecturer in Innovation Studies at the University of East London, Tony founded and led two international development agencies working in Central America and Southern Africa. After he stood down as CEO of Computer Aid International, he completed a PhD in the use of digital technologies in international development (University of London). After one year as a Research Fellow at the United Nations University in Macao, China, Tony joined IDS, where his research focuses on digital development and digital rights. He co-founded the African Digital Rights Network, for whom he has edited reports, including 'Mapping the Supply of Surveillance Technologies to Africa'. He is the series editor of the Zed Books Digital Africa series and co-editor of the collected edition books *Digital Citizenship in Africa*, *Digital Disinformation in Africa* and *Digital Surveillance in Africa*.

Alette Schoon is a Senior Lecturer at the School of Journalism and Media Studies at Rhodes University in South Africa, where she teaches Documentary Filmmaking. Alette worked for many years as a filmmaker, producing hundreds of short films and documentaries broadcast on national television. In the early 2000s she joined Rhodes University and completed a PhD examining mobile phone practices and the digital ecologies of hip hop artists. She has published several journal articles and book chapters on digital cultures, mobile communications and social media in South Africa. Alette has a particular interest in digital and computational methods and organizes an annual winter school called *DigiMethods Africa* for media researchers. She is an active member of the Digital Humanities Association of South Africa. She still produces films, and recently she and her collaborator Dr Hleze Kunju won the NIHSS award for their film *Intellectual Giants of the Eastern Cape*.

Dércio Tsandzana is a lecturer at Eduardo Mondlane University and a Mozambican PhD in Political Science at Sciences Po (France). Since 2013, he has been working on youth, internet, social media and political participation in Mozambique, including online activism for national and international organizations. Alongside his PhD, Tsandzana undertook research work focusing on digital rights and data privacy with Freedom House, Meta/Facebook, Internews/USAID, UNESCO and the Centre for Human Rights Faculty of Law/University of Pretoria. He is a member of the African Digital Rights Network and the African Declaration on Internet Rights and Freedoms Initiative. His latest book (2023) is entitled *Digital Citizenship*.

Marion Walton is an Associate Professor in the Centre for Film and Media Studies at the University of Cape Town, where she has integrated critical data studies into media studies, journalism and informatics curricula. She is also a member of the Data Carpentry community of instructors, who work to teach foundational computational and data science skills to researchers. Marion's research focuses on developing methods, software tools and ethical approaches for the study of digital and social media in South Africa and the Global South. She has published about the development of digital and mobile media in South Africa, with a particular focus on digital inequalities, research ethics, young people's identities, literacies, and their creative and social agency.

ACKNOWLEDGEMENTS

This book is a product of the African Digital Rights Network (ADRN), a virtual association of seventy-five African activists, analysts and academics from over thirty African countries. We were fortunate to be among the founding members of ADRN in the company of African digital rights luminaries such as Juliet Nanfuka, Thobekile Matimbe and Nanjala Nyabola. This book could not have been written without the inspiration of their prior work organizing feminist digital rights citizenship. The wider book series would not be possible without the wealth of knowledge and experience brought together by ADRN's growing membership. We are also grateful to Nick Wolterman and Nadine Staes-Polet for their support at Bloomsbury steering the book through to publication and would like to thank the anonymous external reviewers for the constructive feedback at key stages of the project. Above all, this book exists because of the feminist activists weaving their tapestry of digital citizenship across the continent. It is their vision and agency that sustains us.

Chapter 1

THEORIZING FEMINIST DIGITAL CITIZENSHIP IN AFRICA

Tanja Bosch and Tony Roberts

Feminist digital citizenship is the use of technologies such as smartphones and social media to engage in civic and political life in ways that advance gender rights and justice. Despite efforts to confine them to the private realm, African feminists have a long history of civil engagement in the public sphere to make rights claims and demand justice. However, some gender issues, including sexual rights, reproductive freedoms and bodily autonomy, have remined censored or repressed. Now, a remarkable new wave of active feminist citizenship is unfolding across the African continent, using new information and communications technologies to foreground repressed voices and taboo issues. This book is the first volume authored by African feminists that documents this new wave of feminist digital citizenship and analyses episodes of contention from across the continent's main geographic and linguistic regions.

Social media, for example, has provided feminists with new ways to build safe spaces, to produce powerful multi-media content and to message millions around the world, instantly, repeatedly and affordably. These technical affordances of digital technologies have provided feminists with new action possibilities, including the ability to circumvent the patriarchal editorial control of mainstream media and party-political leadership that have blocked feminist voices and concerns. Social media has afforded feminists novel opportunities to build spaces of mutual aid and self-care, as well as the tools to create influential multi-media content to share with millions around the world in viral campaigns of rights claiming. These digital spaces are not, however, free from power dynamics or gender-based violence, as the chapters that follow illustrate.

One of the most influential campaigns of feminist digital citizenship in Africa was the hashtag campaign #BringBackOurGirls, which erupted across social media in April 2014. Led by Nigerian feminists, the digital campaign ignited a global outcry against the abduction of 276 schoolgirls by Boko Haram in Nigeria and forced government action. What began as a local plea for action was quickly transformed into an international citizens' movement, driven by the power of feminists using digital tools and skills. The hashtag went viral and became a rallying cry for justice, breaking through to mainstream media, securing the support of world leaders and celebrities around the world, and placing immense pressure on the Nigerian government to respond. Yet, it also highlighted the digital divide, as many of the most vulnerable remained unheard in the clamour for change.

Similarly, in 2020, the hashtag #AmINext gained momentum in South Africa, capturing the collective outrage and fear surrounding the epidemic of gender-based violence (GBV). Sparked by the brutal murder of Uyinene Mrwetyana, a young university student, the hashtag became a powerful symbol of the pervasive threat women face daily. #AmINext resonated deeply across South Africa, as women shared their stories of harassment, abuse and fear, demanding accountability and systemic change. This movement of digital citizens forced the country to confront the GBV ignored by mainstream media and produced widespread support for action against gender-based violence, rape culture and femicide.

As the examples featured in this book show, there is a new generation of African feminists making productive use of digital tools and online platforms to circumvent the institutions that (re)produce structural gender (dis)advantage. Social media is being used to evade the patriarchal media and that of legacy political parties and create a new counter-public able to challenge orthodoxies that uphold gender inequities. Using new technological media, feminist digital citizens are able to directly articulate their unfiltered demands for gender rights and justice without establishment editorial control or mediation. Some of these campaigns of feminist digital citizenship have gone viral, reaching nationwide and international audiences, and have secured real changes in policy and practice. Although there remains a gender digital divide limiting access to this digital citizenship, and while it is true that digital platforms themselves are also sites of gender-based violence, the chapters in this book document how digital tools and platforms have become key spaces of feminist agency and digital citizenship, particularly in Africa.

The subtitle of this book is 'Agency, Rights and Resistance', key themes that run throughout this volume. The case study chapters highlight what we call the 'digital dialectic': the contention of opposing uses of digital technologies and their resolution in feminist movements. On the one hand, chapter authors document distinct episodes from across the continent that make positive use of digital technologies to expand feminist agency, rights and resistance. On the other hand, chapters analyse the negative use of digital technology by patriarchal and misogynist actors to conduct gender-based violence and otherwise restrict gender rights and freedoms. The outcome of this digital dialectic is not certain. While the technology provides powerful tools for feminists to actively engage in civic and political life, it is often a double-edged sword, enabling surveillance, disinformation and coordinated campaigns of gender-based violence. This dialectical tension underscores the need for a critical approach to digital spaces, recognizing both their potential to advance gender justice and the risks they pose in amplifying gender-based violence and infringing on rights and justice.

Among the emerging literature on feminist digital activism, this collected edition is unique in several respects: each chapter analyses a case study from a different African country; each case study is authored by a feminist with deep knowledge of that country; case studies are drawn from each of the continent's main geographical and linguistic areas; and all of the chapters use the conceptual lens of African feminist digital citizenship. This book builds upon our previous book, *Digital Citizenship in Africa: Technologies of Agency and Repression*. It was inspired by a single chapter in that previous book called 'Feminist Digital Citizenship in Nigeria' by Sandra Ajaja. This book expands the analysis of feminist digital citizenship in Nigeria to also include Zambia, Botswana, Rwanda, Uganda, South Africa, Guinea, Egypt, Mozambique, Malawi, Sudan and Ethiopia. Authors in this volume adopt an intersectional perspective, highlighting how various dimensions of (dis)advantage, such as gender, race and sexuality, intersect and impact experience. This volume includes a focus on LGBTQ+ struggles, although our next collaboration will be a collected edition dedicated to 'Queer Digital Citizenship in Africa'.

A book titled *Feminist Digital Citizenship in Africa* must address at least four interrelated questions: the feminist question, the digital question, the citizenship question and the Africa question. In this introductory chapter, we address each in turn before the case study chapters expand upon those questions and examine them in more detail. The feminist questions include which kinds of feminisms are explicit or implicit in the specific cases of digital citizenship featured in this book and why. The digital questions include addressing why African feminists use digital tools and online spaces, and what new action possibilities mobile and internet technologies afford to African feminists that were not previously available. The citizenship questions include why we use the framework of citizenship over a framework of activism or feminism alone and why this new wave of feminist rights claiming and demands for accountability is critical to gender justice. The Africa questions include reflection on what is distinctive about feminist digital citizenship on the continent and what it has to contribute to citizenship informed by other standpoints. All of these questions are addressed initially in a brief review of the existing literature and then in greater detail by the authors in the country chapters that follow this introduction.

The remaining sections of this introductory chapter are organized as follows: the next section provides a historical background to feminist citizenship in Africa prior to the digital age to highlight relevant (dis)continuities. A review is then conducted of the existing literature on digital citizenship and feminist digital citizenship, including cyberfeminism, datafeminism and hashtag feminism. A critique of white Western feminism then precedes a review of African feminist digital citizenship to date. To complete the conceptual framework used in this volume, we also review the concept of the 'affordances' of digital technologies and the conformist-reformist-transformist framework of feminist action. This chapter then concludes with a brief overview of the contents of the main case study chapters.

Background

The struggle for gender justice in Africa is at least as old as colonialism (Mama, 2002; Horn, 2025). African feminists were entreated to postpone demands for gender equality first until after independence, then until after the establishment of multi-party democracy, and more recently until after the fulfilment of basic needs. Although the history of feminist citizenship is insufficiently recorded, the Aba Women's Riots are among those that are documented. The injustice of colonial taxes one hundred years ago led to mass protests of thousands of Igbo women in 1929. During the independence struggle in Kenya, women played a central part in the Mau Mau movement, but after independence in 1963, they were urged to prioritize nation-building ahead of gender justice (Nzomo, 2003), as was the case in Zimbabwe, where women played a key role in the liberation struggle but were excluded from power after liberation. In South Africa, 20,000 women marched against laws passed in 1956 and confronted the forces of apartheid. Women repeatedly led collective action outside of the mainstream political parties that were controlled by male leaders. In Kenya, the Green Belt Movement empowered women in the 1970s to combat deforestation and promote sustainable land use. Around the turn of the millennium, the Women of Liberia Mass Action for Peace successfully used nonviolent mass protests and sit-ins to help end Liberia's civil war. This legacy of feminist engagement in civil and political life provides the inspiration and foundation for feminist digital citizenship in Africa and is not disconnected from it. Majority world feminisms have been deeply influenced by the legacies of colonialism and patriarchal structures of violence, injustice and inequality. Women and gender-diverse people in the majority world face intersecting forms of oppression, including poverty, militarism, environmental degradation, religious fundamentalism and neoliberal globalization.

Feminist digital citizenship began to emerge in the early 2000s, coinciding with the rapid expansion of internet access and mobile technology across the continent. Initially, digital advocacy and mobilization were peripheral to existing methods, but the affordances of text messaging (Ekine, 2010), blogging and social media enabled new issues to be tabled and previously marginalized feminist voices to be heard (Nyabola, 2018). While uneven access to digital devices, connectivity and digital literacies continue to structure hierarchies of digital access (Hernandez and Roberts, 2018), it remains the case that some African feminists have been able to leverage digital technologies to create spaces where they can assert their voices and agency, claim rights and resist long-standing patriarchal structures (Mendes, Ringrose and Keller, 2019).

As more African women came online in the 2010s, they began to carve out spaces for dialogue, activism and community building, challenging traditional gender norms and pushing for greater inclusion and representation in the digital realm. The #BringBackOurGirls campaign became a global rallying cry, highlighting the power of digital spaces to mobilize support and draw attention to issues of gender-based violence in Africa (Ogbonna, 2018; Agunwa, 2023). Over time, African feminism evolved, adapting to changing socio-political contexts and embracing digital technologies as a tool for activism and organizing (Mendes, Ringrose and Keller, 2018; Clark and Mohamed, 2023; Horn, 2025).

The central research question that this book addresses is, how do African feminists leverage digital platforms to engage in civic and political life in ways that expand agency, rights and resistance to advance gender equality and social justice? Our study explores the role of feminist digital citizenship in Africa, using case studies to document a range of diverse and intersectional experiences across Anglophone, Francophone and Lusophone Africa. While the book seeks to foreground diverse expressions of feminist digital citizenship in Africa, the study is situated in the context of a host of structural challenges that include the use of digital technologies to limit the space for digital citizenship, including the gender digital divide, digital surveillance and profiling, disinformation, cyber harassment and online gender-based violence.

Literature review

This section reviews the existing literature on citizenship, digital citizenship, feminist digital citizenship and African feminist digital citizenship to provide a framework for the analysis contained in the book. The literature on citizenship and digital citizenship is more fully reviewed in our previous volume (Roberts and Bosch, 2023). Here we focus in greater depth on feminist digital citizenship and African feminist digital citizenship as well as on the affordances of digital technology for feminist citizenship and present a framework to differentiate conformist, reformist and transformist iterations of feminist digital citizenship.

Citizenship

In general speech, 'citizenship' is often associated with a status bestowed (or withheld) by a relatively powerful state on relatively passive individuals – and is often certified by a passport or national ID card. However, in citizenship studies, the term 'citizenship' can also refer to the active engagement of any person in the civic and political life of a community. From this perspective, citizenship is an activity that anyone can engage in and is not a status to be bestowed or certified. Such citizenship may or may not be overtly political; it can take the form of participation in a school board or neighbourhood association, or an advocacy or self-help group. In this book, citizenship is understood as the active process of engaging in the civic and political life of a community, which includes forms of exercising, defending and claiming rights. Nyamu-Musembi (2006) argues for this agency-based understanding of citizenship, pointing out that such 'actor-orientated perspectives are based on the recognition that rights are shaped through actual struggles informed by people's own understandings of what they are entitled to'.

Digital citizenship

Digital citizenship, put simply, is the process of participating in civic or political life using digital tools or online spaces (Roberts and Bosch, 2023). As aspects of social, economic and political life increasingly take place online, so does the way that people engage in civic and political life. Parent–teacher association meetings can take place online, shared interest groups may coordinate using instant messaging apps like WhatsApp, and party-political campaigning may take place on social media. Mainstream media (newspapers, radio and TV stations) have remained male dominated, but digital media platforms have provided a place where it is possible – for those with the requisite digital devices, connectivity and digital literacies – to create counter-publics and counter-narratives by escaping the editorial control of the patriarchs in legacy media. Our previous book explored eight case studies illustrating how citizens have made creative use of digital technologies to claim rights, hold government accountable and cope with the limitations and restrictions of digital citizenship (Roberts and Bosch, 2023). It is always important to note that access to digital tools, connectivity and digital literacies is unevenly distributed between countries and within countries along intersectional dimensions of power and privilege, including gender, 'race', coloniality, class and disability (Mutula, 2008; Benjamin, 2018; D'Ignazio and Klein, 2023). This means that digital citizenship is rarely the work of a representative cross sample of the whole population. More often, urban, educated and relatively wealthy voices are over-represented on digital platforms and otherwise marginalized communities are usually under-represented, leading to the criticism that digital platforms often (re)produce existing patterns of (dis)advantage (Hernandez and Roberts, 2018).

Affordances of digital technologies

Affordance theory has proven to be useful in analysing the distinctive benefits of different technologies. Specifically, it is a way to analyse what new action possibilities a specific technology enables or constrains (Gibson, 1977; Norman, 1988; Roberts, 2021). For example, we can analyse what it is about a particular digital tool or space that affords feminists the possibility to create 'safe spaces', 'space of care' or 'organizing spaces' in which mutual aid, solidarity and mobilization can take place – and why (Currier and Moreau, 2016; Mhiripiri and Moyo in Mutsvairo, 2016). The concept has been used to analyse the use of Twitter (now X) in the Arab uprising in Tunisia and Egypt (Tufekci, 2017) and the use of internet shutdowns to stifle protest (Anthonio and Roberts, 2023). For the increasing numbers of people who have digital tools, connectivity and digital literacy, the world wide web and social media afford citizens the ability to access information and to publish their thoughts and demands instantly, interactively, repeatedly, nationally and globally, using text, images and video. These new digital affordances have resulted in previously underheard social groups and issues becoming national and global news, including the #MeToo campaign that created global awareness of the extent of sexual abuse and exploitation and the jailing of many perpetrators. However, it is important to recognize that the same digital tools that afford these positive possibilities also afford the negative possibilities to conduct surveillance, disseminate disinformation and conduct GBV, digital misogyny or transphobia.

Feminist digital citizenship

Feminist digital citizenship is the use of digital tools, like mobile phones and social media, by feminists to play a full part in civic and political life, including the making of rights claims for gender justice. In recent years, we have seen a wide range of feminist engagements 'that emerge from the interface of digital platforms and activism today' (Baer, 2016), with some claiming that the use of digital technologies is one of the defining features of fourth-wave feminism from 2012 onwards (Munro 2013). Whether it is a defining feature of a new wave of feminism or not, what is clear is that the use of mobile and internet technologies has helped create a global community of feminists across the African diaspora who use digital technologies for discussion, debate and activism (Akinbobola, 2018; Sobande, 2020).

Feminist critiques of technology emerged in the 1980s, including from the work of Cynthia Cockburn (1985), Judy Wajcman's techno-feminism (1985; 1991; 2004) and Donna Haraway's (1985; 1991) cyberfeminism, which examined whether technology was ever gender-neutral and what use feminists might be able to make of new technologies. Donna Haraway's (1991) 'A Cyborg Manifesto' explored the relationship between cyberspace, digital media technologies and gender. Haraway's cyberfeminism has been periodically revived and reworked by different feminist scholars and is used by authors in this volume to analyse feminist digital citizenship in Mozambique and Malawi (Kennedy, 2000; Kember, 2002; Paasonen, 2011).

Especially since the turn of the millennium, the use of digital technologies by feminists has become a focus of scholarly attention (Clark, 2014; Hester, 2016; Fotopoulou, 2016; Keller, 2018). In their book *Digital Feminist Activism*, Mendes, Ringrose and Keller (2019) extensively document powerful examples of how feminist activists can leverage the affordances of digital technologies to build solidarity, tackle rape culture and engage in rights-claiming citizenship. Although their examples are primarily from the Global North, their book is valuable for showing how online spaces (re)produce gender power structures from offline spaces that (dis)advantage some groups over others and how they exist even among feminist activists shaping whose voices are heard and whose interests are prioritized.

In their paper 'Digital Citizenship in a Global Society: A Feminist Approach', Henry, Vasil and Witt (2022) explore digital citizenship through the lens of feminist theory, highlighting how digital spaces can both empower and oppress individuals, particularly women and marginalized groups. The authors argue that while digital technologies offer opportunities for participation and advocacy, they also reproduce existing inequalities related to gender, race and class. The authors emphasize the importance of creating inclusive digital environments where diverse voices can be heard without fear of harassment or discrimination. By applying feminist principles, such as intersectionality and collective action, the authors advocate for a more equitable form of digital citizenship that challenges power imbalances and promotes justice in the global digital landscape.

In *Data Feminism*, D'Ignazio and Klein (2023) explore feminist research methods. The book critiques traditional practices in data science, highlighting how they often perpetuate systemic inequalities, particularly those related to gender, race and power. D'Ignazio and Klein draw on feminist theory to propose a framework that prioritizes justice, accountability and the amplification of marginalized voices in the collection, analysis and presentation of data. Through case studies and examples, they emphasize the need for ethical data practices that challenge existing power structures and create more equitable outcomes in society. The book is a call for rethinking the ways data is used, and it encourages the development of a more inclusive data science that actively works to dismantle bias and inequity.

Hashtag feminism is the practice of using hashtags like #MeToo and #YesAllWomen to create online rallying points for feminist mobilization and counter-narratives. Hashtag feminism has emerged as a powerful tool in contemporary digital citizenship, employing the accessibility and reach affordances of social media to challenge gender injustice and amplify marginalized voices. Lorza (2021) discusses how hashtag feminism, through campaigns like #MeToo and #YesAllWomen, has created digital spaces where individuals can share personal experiences of GBV and discrimination, fostering a sense of solidarity across global communities. Hashtag campaigns can transcend national borders and cultural contexts, making feminist discourse more accessible to a broader audience. Dixon (2014) explores the intersection of hashtag feminism and mainstream media, noting that hashtags can focus global attention on specific causes, attracting media coverage that can lead to political influence and policy action. Chen (2018) highlights the dual nature of hashtag activism, noting that while hashtags can elevate feminist issues into public consciousness, they are also susceptible to commercial exploitation or trolling, which can undermine their effectiveness. Despite these challenges, hashtag feminism remains a vital component of feminist digital citizenship, providing a decentralized, inclusive and far-reaching approach to addressing gender injustice.

The critique of white Western feminism

The existing literature on digital feminism has been dominated by studies of the West, by Western researchers, that rely primarily upon Western conceptual frames. This volume aims to make a modest contribution towards redressing this epistemic violence. It provides ten case studies of episodes of feminist digital citizenship from ten different African countries that are authored by scholars with deep contextual knowledge of those countries. The chapters foreground scholarship by other African researchers of digital feminism.

Many majority world feminists have critiqued mainstream Western feminism for its narrow focus on issues relevant primarily to white, middle-class women. The feminist movement in the West has come under critique for its failure to 'recognize the different historical experience of black women compared to that of white women and has been aggressive towards their cultural values and struggles for freedom as black women' (Roberts, 1983: 175). Kendall (2020), for example, argues that mainstream feminists do not focus on meeting basic needs (e.g. food insecurity, housing). She describes mainstream feminism as 'white feminism' due to its erasure of issues that affect majority world communities and, in response, highlights an intersectional approach to feminism. Similarly,

Chandra Mohanty (1991) has questioned the canonized feminism of the Global North and argues that Western feminist scholarship tends to homogenize majority world women, positioning Western women as liberated and majority world women as oppressed. She argues that this approach reinforces colonialist power dynamics, emphasizes the importance of intersectionality and advocates for an understanding of women as active agents, as we see in the case studies outlined in this volume.

Hashtag feminism has been criticized as superficial 'slactivism', incapable of addressing the intersecting forms of oppression faced by women of colour and queer and transgender individuals. Such criticisms are often accompanied by a call for more inclusive and intersectional approaches to feminist activism. For example, Rottenberg (2019) points out that while campaigns like #MeToo have been useful as a mobilization tool and a springboard for 'significant political or even structural change' (ibid: 41), not only was the campaign focused on white, celebrity-status women, but it was an individualizing campaign, not involving 'collective demands for systematic change' (ibid: 45). Rottenberg argues, 'by neglecting any kind of structural critique, the movement elides the women who are perhaps most vulnerable to violence – sexual or otherwise – such as immigrants, domestic workers, and low-income women of colour' (ibid: 45).

Critics of hashtag feminism highlight its centring of white Western voices. Trott, for example, argues that these campaigns 'obscure the absence and recognition of marginalised women and those who are already more vulnerable in regards to experiencing sexual assault' (Trott, 2021: 1126). Zarkov and Davis (2018) question whether hashtag feminism advances or provides an obstacle to wider social mobilization. The marginalization of African experience in white Western digital feminism is a form of epistemic violence that silences and erases African women's knowledge and lived experience. This volume provides a space for the analysis of African feminist voices. By providing a series of rich case studies of feminist digital citizenship in Africa, we hope to provide foundations that other scholars will correct, improve and extend.

African feminist digital citizenship

Despite these critiques and failings, African women and non-binary individuals are increasingly using digital technology and social media platforms to amplify their voices, build supportive communities, and claim rights and gender justice. They harness these tools to challenge oppressive systems, share their experiences and create spaces for empowerment and activism. Feminist digital citizenship in Africa has emerged as a powerful tool for amplifying marginalized voices, challenging dominant narratives and exploring issues of injustice. This volume presents examples of African feminist digital citizenship from around the continent, highlighting their agency, rights and resistance.

The last decade has witnessed the documentation of a specifically African literature on feminist digital citizenship, but its roots go back much further.

Ghanaian feminist Nana Darkoa Sekyiamah started her blog about sexuality and sexual health in 2009. Minna Salami (2014) reviewed this blog among other notable examples of early African digital feminism in her article on GenderIT.org about the 'coming of age' of African digital feminism in the wake of – and in opposition to – fourth-wave white feminism hashtags including #EverydaySexism and #Fem2.0. McLean and Mugo (2015) wrote a seminal article one year later asking, 'How can digital spaces make possible a feminist future for the queer African woman?' Njoroge (2016) examines what was perhaps the global breakthrough for African digital feminist citizenship – the #BringBackOurGirls campaign, which leveraged digital tools and the global reach of the African digital diaspora to raise international awareness about the kidnapping of Nigerian schoolgirls by Boko Haram.

The work of Zimbabwean activist scholar Sibongile Mpofu (2016) examines how feminists use online spaces and tools to engage in civic and political life. Nanjala Nyabola's (2018) extensive publications analyse key case studies of digital feminism in Kenya and the wider region, including the use of blogging and social media as forms of resistance. Neema Iyer (2022), from the Ugandan research agency Pollicy, is among those feminist scholars who have noted that despite the benefits afforded by digital technologies, online spaces often embody and reproduce existing (offline) systems of oppression and GBV. Her work is notable for recommending in response a radical shift to develop alternative digital networks, feminist methodologies and decolonial practices.

This book was inspired in part by the chapter on Feminist Digital Citizenship in Africa that Sandra Ajaja (2023) contributed to our previous book on digital citizenship (Roberts and Bosch, 2023). Ajaja identified the need for a focused study on the feminist digital citizenship across the continent, and her original chapter advanced those discussions by exploring the intersection of gender, technology and political engagement across multiple cases. Although it was not solely dedicated to African case studies, the most expansive and in-depth study of feminist digital citizenship to date was perhaps the excellent *African Women in Digital Spaces*, edited by Kibona Clark and Mohammed (2023). That volume provides a rich collection of essays on how African women – in different countries and on different continents – navigate and influence digital environments. Together, these works demonstrate the diverse and evolving nature of feminist digital citizenship across Africa while also emphasizing the ongoing challenges of navigating digital spaces in which offline power relationship are never entirely absent.

As Horn (2025: xi) notes, 'for most of the history of African feminism has been the history of cis-gendered women and girls', but there is also an emerging literature on queer digital feminism in Africa, a subject on which we plan to dedicate a future collected edition in this series. A notable example of scholarship in this neglected space is provided by Mhiripiri and Mpofu (2016), who document how the organization Gays and Lesbians of Zimbabwe (GALZ) utilizes Facebook as a platform to create an alternative public sphere for LGBTQ+ advocacy in a society rife with homophobia. Through digital platforms, GALZ and similar groups can engage in discussions and mobilize support in ways that might otherwise be impossible in more restrictive offline environments.

As this present collection highlights, digital platforms can afford marginalized groups the ability to create safe spaces, build collective agency, resist oppression and claim rights. However, this potential comes with significant risks. Currier and Moreau (2016) warn that while digital media can empower activists, it can also create a focus for homophobia. They caution that 'digital media can fuel political homophobia targeting LGBTI activism' (p. 234) and note that although digital activism may elicit engagement from domestic governments, activists may lose control over the narrative as their movements gain domestic and transnational visibility (p. 243). This digital dialectic – the tension between affordances for agency and repression – underscores the complex dynamics of digital activism in Africa, where the benefits of visibility often come at the cost of heightened risks.

Transformational digital citizenship in Africa

Given the diversity of digital feminisms and tactics of digital feminism reviewed, we find it useful to study this variance using an analytical typology. What feminist practices and digital citizenship approaches are available and appropriate in difference contexts and situations? One useful approach for thinking this though borrows from the work of Zambian feminist Sara Longwe and South African feminist researcher Ineke Buskens.

In feminist theory, Maxine Molyneux (1985) made an important analytical distinction between women's 'practical gender interests' and their 'strategic gender interests'. From this perspective, women's immediate practical needs include income, childcare and access to safe spaces. Without diminishing the foundational importance of attending to women's practical needs, Molyneux contrasted them with what she termed women's *strategic interests*. Women's strategic interests, she argues, include ending the gendered division of labour, male violence against women and male domination of politics and the economy. Molyneux argued that the immediate practical problems that women experience are symptoms caused by these deeper structural issues of unequal power and control, and the symptoms would remain until the root causes were addressed. In some ways, this analysis echoes Rosa Luxembourg's (1908) earlier polemic 'Reform or Revolution', in which she argued that the action of civil society organizations to reform capitalism only prolong the structural basis of exploitation and that ending it required more transformational change.

Zambian scholar Sara Longwe (in Young, 1993) critiqued what she regarded as the artificial separation of practical and strategic interests, arguing that action to attend to 'women's practical needs' often impacts 'strategic gender interests'. Longwe argued that interventions should target *both* practical needs *and* strategic interests by using reflective discussion groups to build gender consciousness among those experiencing gender injustice in order to inform their own collective action to overcome it. Young (1993: 156) argued for the need to build 'transformational potential' by organizing opportunities for women's collective dialogue and group analysis to translate concerns about their immediate practical interests into collective action that addressed unequal gender power relations.

South African gender scholar Buskens (2010) devised an expanded threefold framework for analysing gender research. She argued that there are three categories of gender research: research with conformist intent, reformist intent or transformist intent.

Conformist gender research intends to record and analyse the practical consequences of gender inequity and help people to be resilient in the face of an adverse and unjust reality. Documenting women's immediate and practical needs is of critical importance. However, this kind of research does not intend to reform the gendered social norms or structural power relationships that produce gender inequality. Buskens argued that this research consciously or unconsciously facilitates conforming with the status quo: helping people to better cope with existing unequal gender relations without setting out to challenge or change them.

Reformist gender research is research that intends to raise awareness about unequal gender norms, practices or legislation in order to reform and improve them. Research in this category aims to identify and draw attention to gender norms and values that are discriminatory in order to modify and change those gender norms and values. It does not, however, intentionally set out to identify and overcome the structural patriarchal and intersectional power relationships that give rise to unequal gender norms and values.

Transformist gender research is research that intentionally sets out to analyse and transform the unequal power structures that give rise to and sustain gender injustice and thereby (re)produce and structure unequal gender relations. Transformist gender research is the only type of research that is intent on identifying the root causes of gender injustice and overcoming them.

It is important to note that we do not claim there is a hierarchy of importance between conformist, reformist and transformist feminist digital citizenship. There are compelling reasons to attend to the immediate and practical needs of people negatively affected by gender injustice. There is also good reason to focus on reforming unjust laws and gender norms. However, if all available time and energy is directed exclusively towards firefighting immediate and practical needs, then subsequent generations will necessarily face the same structural problems of gender inequity and injustice. It may be necessary and prudent to start with conformist initiatives and to view reformist activities as a necessary step. They may indeed be understood as essential first steps. The only claim made here is that transformist work is also essential if we are to tackle the root causes of structural gender injustice.

Roberts (2015; 2016) used these categories not to evaluate gender research as Buskens had but to evaluate projects and programmes using digital technologies for social change. Later Faith, Roberts and Berdou (2018) used the conformist, reformist and transformist framework to evaluate digital feminist initiatives designed to tackle gender inequity in workplaces. In this chapter, we propose that the framework has utility for analysing different types of feminist digital citizenship in Africa.

We suggest that it is useful analytically to distinguish between different approaches of feminist digital citizenship in Africa – specifically between those designed to (a) help people to cope practically with gender injustice as it is, (b) understand how to reform unequal gendered social norms and values, and (c) transform the structural power relations that cause gender injustice. Using this approach, we suggest that there are three categories that we can use to analyse the types of feminist digital citizenship featured in this book and elsewhere:

Conformist feminist digital citizenship describes feminist citizenship practices where digital mobile online spaces intend to address the immediate and practical needs of those experiencing gender injustice. This might involve providing safe online spaces, mutual aid and care, training and support. Conformist practices are designed to help people to cope with or be resilient in the face of existing gendered disadvantage. These practices could help people experiencing gendered exclusion to learn new skills to survive or compete despite discrimination. Conformist feminist practices do not intentionally set out to reform laws, change gendered social norms or tackle underlying gender power relationships.

Reformist feminist digital citizenship refers to feminist initiatives that intend to reform gendered social norms and values using digital technologies or online spaces. Reformist feminist digital citizenship might, for example, involve a hashtag campaign like #MeToo conducted on social media to call out GBV and change social norms around (un)acceptable behaviour. Such initiatives intend to shame perpetrators, reform laws and shift patterns of behaviour. It can be argued that we have very good reason to value such initiatives. At the same time, if initiatives do not set out to tackle the structural power relations of heteropatriarchy that give rise to, and reproduce, gender injustice, they will remain in place for future generations to suffer.

Transformist feminist digital citizenship describes initiatives that set out to identify the structural root causes of gender injustice and guide action to fundamentally transform the situation and create gender justice. If it is accepted that gender injustice is determined by intersecting power structures (Crenshaw, 1989; hooks, 2000; Hill-Collins, 1990, then understanding which power structures are causal in a specific context is logically necessary to accurately guide action to uproot the causes of gender inequity. This is the most challenging level of feminist digital citizenship and requires more time and resources and wider coalitions. Transforming the matrix of related power structures is not a task that can be achieved by an individual, by a single project, or in one year.

In summary, transformist feminist digital citizenship is an approach that sets out to identify the intersectional power relationships that are structuring gender injustice in a given context in order to inform action to dismantle those systems of oppression and secure gender justice.

The chapters in this book illustrate how ethno-religious, colonial and heteropatriarchal power structures, among others, uphold gender injustice in ways arguably unique to Africa and which would be largely inaccessible to researchers who lack deep contextual knowledge of each country's political and cultural specificity. It is for that reason, we argue, it is important to provide platforms for African researchers to conduct their own analyses in publications such as this.

In the next section, we provide an overview of the country case study chapters, including comments on how they contribute to the distinct elements of the conformist, reformist, transformist framework.

Outline of chapters

The chapters in this volume present a range of case studies and reflections to extend our understanding of feminist digital citizenship. Each chapter highlights a distinct episode of feminist digital citizenship in a different country using a range of methodological and conceptual approaches. All of the chapters contribute to answering the overall research question of how African feminists use digital technologies to engage in civic and political life in ways that expand agency, rights and resistance to advance gender equality and social justice. However, each author introduces a more specific question tied to a particular case study and context. This section summarizes the chapters and approaches taken to answering the question.

In **Chapter 2**, Ochega Ataguba analyses the rise of Nigeria's feminist digital citizenship on Elon Musk's micro-blogging site X (formerly Twitter). The chapter uses the #UjuAnya online debates as a case study. Uju Anya, a well-known queer feminist in Nigeria, used the death of Britain's Queen Elizabeth to critique the country's colonial legacy and the Biafra genocide under her rule. Much of the backlash against the initial post took the form of gender-based attacks on Uju Anya and her sexuality. The chapter author, Ochega Ataguba, skilfully analyses how digital feminists have created online safe spaces to provide refuge and respite care when facing online gender-based violence (a form of conformist digital citizenship to meet people's immediate and practical gender needs) as well as using social media like X/Twitter to challenge heteropatriarchal norms explicitly online (a form of reformist digital citizenship to shift social norms). The author concludes that while this situation is 'a far cry from revolutionary', the affordances of digital spaces provide new action possibilities for dissidence and represent a clear improvement over the alternatives.

In **Chapter 3**, Chishimba Kasanga and Priscilla Boshoff explore the rise and role of the Zambian Feminists Facebook page, which was founded in 2018 as a platform for Zambian women to challenge patriarchy and engage in feminist activism in a predominantly conservative and heteronormative society. The page serves as a digital space where GBV, LGBTQ+ rights and various patriarchal practices are discussed. It also provides Zambian women with a sense of community and belonging, as expressed in its tagline, 'of course we are angry'.

Using digital ethnography, content analysis and interviews, the authors investigate how women engage with and interpret the feminist content shared on the platform. The chapter highlights the complexities of feminist activism in Zambia, including the role of the page administrator as a curator and moderator, as well as the struggles women face in expressing their feminist identities both online and in their daily lives. The authors emphasize that while the digital space provides an avenue for feminist action, it also carries risks, such as trolling and social backlash. Despite these challenges, the Zambian Feminists Facebook page continues to serve as an important platform for advocating for gender equality and offering support to women navigating patriarchal norms.

In **Chapter 4**, Nyx McLean analyses transgender people's online solidarity strategies for keeping each other safe in Botswana, Rwanda, South Africa and Uganda. The chapter tackles a neglected subject, examining the ways that transgender, non-binary and gender-diverse (TNBGD) people are using digital technologies to actively participate in civic and political life online and challenge GBV. The chapter vividly describes how a range of tactics are employed by TNBGD digital citizens to keep themselves safe when targeted by transphobic individuals and groups. The tactics include the use of closed groups to create a safe space of mutual aid and support (a form of conformist digital citizenship to meet people's immediate and practical gender needs). Other conformist tactics employed include blocking and mass reporting of transphobic individuals to social media platforms.

In **Chapter 5**, Alette Schoon and Marion Walton analyse the issue of unequal access to mobile phones in South Africa's rural Ciskei and women's practices of 'digital frugality', managing connectivity with minimal resources in marginalized settings. The chapter argues for a conception of African feminist digital citizenship that acknowledges women's struggles and their survival strategies. The authors analyse frugal mobile use in two research sites characterized by conditions of 'precarious survival' to illustrate the intersectional disadvantages of low-income Black women in one of the world's most unequal societies. The chapter authors draw attention to the co-existence in South Africa of elite spaces of predominantly white feminist digital citizenship and what they term 'everyday African feminist survivalist digital citizenship' – the frugal use of limited digital access to survive a precarious existence (a form of conformist digital citizenship intent on coping with adversity rather than challenging gendered power structures).

In **Chapter 6**, Clovis Bergère examines the (reformist) feminist digital citizenship campaign #guineennedu21esiecle, launched in Guinea in 2016. The chapter analyses an online and offline campaign by a collective of young Guinean women across Facebook, Twitter, Instagram and YouTube to advocate for women's rights and challenge traditional celebrations of International Women's Day in Conakry. One of its key actions was a 'digital blackout' protest following the death of M'mah Sylla, a young Guinean woman who was a victim of sexual violence and medical mistreatment. By connecting contemporary digital campaigns to this historical lineage, Bergère argues that these movements retain the radical potential of earlier feminist politics while adapting to the new digital landscape, challenging the neoliberal framing of digital technology as merely innovative or disruptive.

In **Chapter 7**, Manal Hassan examines (transformist) moments in feminist digital citizenship during the Egyptian revolution. The author analyses how the affordances of specific digital tools and online spaces enabled Egyptian feminists to enact new forms of feminist citizenship that had been denied them prior to the revolution in 2011. The occupation of Tahrir Square created a physical space where women were able to openly engage in political debate without GBV. Although this physical space was violently closed down by the military, activists promptly opened new online spaces for feminist digital citizenship as an expression of fourth-wave feminism. The author makes extensive use of affordance theory to analyse why different digital technologies enabled forms of feminist citizenship that were impossible to safely enact in the offline spaces in Egypt. FemiHub is an interesting example where an online community was built to create a safe, supportive space for young women to nurture feminist solidarity (a conformist activity) but which developed into a space of consciousness-raising from which they orchestrated active engagement in online and offline policy debates (a reformist activity).

In **Chapter 8**, Dércio Tsandzana and Lissungu Mazula explore the emergence of feminist digital activism in Mozambique. They highlight how the use of digital platforms has empowered women to challenge traditional norms, patriarchal structures and GBV while also emphasizing the historical roots of feminism in the country, dating back to the National Liberation Struggle led by FRELIMO. The Sou Ntavase campaign in 2020 used Facebook and WhatsApp to demand justice for a child survivor of sexual violence and secured legal reforms. The following year, a coalition of NGOs called the Observatório das Mulheres formed to help women make rights claims using digital platforms. While feminist digital citizenship in Mozambique is reformist in intent, the chapter also discusses challenges, including digital violence, limited internet access and cultural resistance to feminist movements.

Chapter 9, by Jones Maweranga and Godwins Lwinga, analyses the feminist digital campaign against Mwiza Chavura's pro-rape song. The song, which describes a man's plan to rape a woman for rejecting his advances, ignited a wave of feminist digital citizenship on platforms such as Facebook and in online tabloids like Nyasa Times and Malawi24. Malawian feminists mobilized online to critique the song and achieve its (reformist) banning. The chapter explores the intersectionality of feminist digital citizenship, cyberfeminism and rape culture, demonstrating how digital tools provided women a space to resist deeply rooted rape culture and demand accountability. By utilizing digital media to amplify their voices, feminists were able to organize widespread resistance against a song that reflected and perpetuated harmful societal norms. This digital feminist activism also led to greater awareness of Malawi's rape culture and influenced public discourse on gender justice.

In **Chapter 10**, Maha Bashri examines how digital platforms are transforming feminist activism in Sudan, particularly in the context of ongoing political upheaval. The chapter focuses on how feminist movements, historically constrained by patriarchal and class structures, have transitioned into the digital realm, offering new possibilities for women to challenge and reform traditional power dynamic and gender norms. Through case studies of two organizations, Shabaka and Andariya, the chapter also explores how Sudanese women use digital tools to contest gendered citizenship and create counter-narratives that challenge patriarchal norms. Despite the potential of digital activism, challenges remain due to the gender digital divide, conflict and internet shutdowns that limit access. Nevertheless, these digital feminist initiatives represent a significant shift in Sudanese feminist movements, offering new ways to engage in political and social discourse.

And finally, in **Chapter 11**, Selamawit Tezera Chaka investigates how digital technologies have been leveraged to advance feminist digital citizenship in Ethiopia despite low internet penetration, gender divides and frequent internet shutdowns. Feminist movements like Yikono and the Yellow Movement have used social media to raise awareness about GBV and women's rights, amplifying their messages in both online and offline spaces. The chapter outlines key campaigns, such as the fight against the tax on menstruation products, and digital activism that followed high-profile cases of GBV, like the Justice For Hanna campaign. While social media has offered a platform for feminist voices, Ethiopia's patriarchal norms, technology-facilitated violence and political instability pose significant setbacks. The chapter also discusses how the war in Tigray has impacted the feminist movement, leading to divisions within the community. Digital activism, despite its limitations, has proven valuable in Ethiopia, offering a space for solidarity, community-building and the amplification of marginalized voices, but it continues to grapple with state censorship and deep-seated misogyny.

Conclusion

This collected edition extends our understanding of feminist digital citizenship in Africa in four important ways. The book provides a unique platform for African feminists to articulate in-depth analysis of digital citizenship in their own countries based on deep contextual understanding of the local political and cultural context. Together, they have created the most extensive range of country case studies of feminist digital citizenship from across the continent, upon which other researchers can now build. The chapters explain why African feminists enthusiastically use social media and online platforms to exercise and expand their digital citizenship. Finally, it offers a novel framework of conformist, reformist and transformist feminist digital citizenship to guide intersectional analysis and action.

The distinct affordances of digital technologies have enabled African digital feminists to circumvent conservative institutions that have long ignored or inadequately addressed gender justice issues and play a fuller role in civic and political life. The digital dialectic affords African feminists new action possibilities to expand their agency, rights claims and resistance. However, digital platforms also afford repressive actors new means of surveillance, disinformation and GBV. How this contestation will be resolved is yet to be determined. African feminists are making productive use of digital tools and spaces to create conformist spaces of safety and self-care, to launch reformist campaigns to shift gender norms and values and to organize and mobilize transformist feminist digital citizenship to uproot the structural power relationships of patriarchy, colonialism and heteronormativity.

The contributions of the authors in this volume and the struggles being fought by feminist digital citizenship across the continent give us reason to hope and celebrate their remarkable agency, rights claiming and resistance.

References

Akinbobola, Y. (2018), 'Defining African Feminism(s) While #BeingFemaleinNigeria', in *African Diaspora*, https://brill.com/view/journals/afdi/12/1-2/article-p64_5.xml

Anthonio, F. and Roberts, T. (2023), 'Internet Shutdowns and Digital Citizenship', in T. Roberts and T. Bosch (eds), *Digital Citizenship in Africa: Technologies of Agency and Repression*, London: Zed Books.

Baer, H. (2016), 'Redoing Feminism: Digital Activism, Body Politics, and Neoliberalism', *Feminist Media Studies*, 16 (1), https://www.tandfonline.com/doi/full/10.1080/14680777.2015.1093070.

Barnes, T. (1999), *We Women Worked so Hard: Gender, Urbanization and Social Reproduction in Colonial Harare, Zimbabwe, 1930-1956*, Portsmouth, NH: Heinemann.

Benjamin, R. (2018), *Race After Technology: Abolitionist Tools for the New Jim Code*, Cambridge: Polity Press.

Buskens, I. (2010), 'Agency and Reflexivity in ICT4D Research: Questioning Women's Options, Poverty, and Human Development', *Information Technologies & International Development*, 6(Special Edition): 19–24.

Chen, G., Pain, P., and Barner, B. (2018), '"Hashtag Feminism": Activism or Slacktivism?', in D. Harp, J. Loke and I. Bachmann (eds), *Feminist Approaches to Media Theory and Research*, Cham, CH: Palgrave MacMillan, 197–218.

Clark-Parsons, R. (2018), 'Building a Digital Girl Army: The Cultivation of Feminist Safe Spaces Online', *New Media and Society*, 20 (6), https://journals.sagepub.com/doi/abs/10.1177/1461444817731919.

Clark-Parsons, R. (2021), '"I See You, I Believe You, I Stand With You":# MeToo and the Performance of Networked Feminist Visibility', *Feminist Media Studies*, 21 (3): 362–80.

Cockburn, C. (1984), *Machinery of Dominance*, London: Pluto Press.

Crenshaw, K. (1989), 'Demarginalizing the Intersection of Race and Sex: A Black Feminist Critique of Antidiscrimination Doctrine, Feminist Theory and Antiracist Politics', *The University of Chicago Legal Forum*, 140, 25–42. Available at: https://scholarship.law.columbia.edu/faculty_scholarship/3007.

Currier, A. and Moreau, J. (2016), 'Digital Strategies and African LGBTI Organizing', in B. Mutsvairo (ed.) *Digital Activism in the Social Media Era: Critical Reflections on Emerging Trends in Sub-Saharan Africa*, Cham, CH: Palgrave MacMillan, 231–47.

D'ignazio, C. and Klein, L. F. (2023), *Data Feminism*, Cambridge, MA: MIT Press.

Dixon, K. (2014), 'Feminist Online Identity: Analyzing the Presence of Hashtag Feminism', *Journal of Arts and Humanities*, 3(7): 34–40.

Ekine, S. (2010), *SMS Uprising: Mobile Activism in Africa*, London: Pambazuka Press.

Emejulu, A. and McGregor, C. (2020), 'Towards and Radical Digital Citizenship in Digital Education', *Critical Studies in Education*, 60 (1): 131–47, https://www.pure.ed.ac.uk/ws/portalfiles/portal/38412413/EmejuliMcGregorCSE2016TowardsARadicalDigitalCitizenship.pdf.

Faith, B., Roberts, T. and Berdou, E. (2018), *Towards a More Gender-Inclusive Open Source Community*. Washington: Digital Impact Alliance.
Fotopoulou, A. (2016), *Feminist Activism and Digital Networks: Between Empowerment and Vulnerability*, London: Palgrave.
Gibson, J. (1977), 'The Theory of Affordances', in R. Shaw and J. Bransford (eds), *Perceiving, Acting, and Knowing*, London: OUP.
Haraway, D. (1988), 'Situated Knowledges: The Science Question in Feminism and the Privilege of Partial Perspective', *Feminist Studies*, 14 (3): 575-99.
Haraway, D. (1991), *A Cyborg Manifesto*, Minneapolis: University of Minnesota Press.
Henry, N., Vasil, S. and Witt, A. (2022), 'Digital Citizenship in a Global Society: A Feminist Approach', *Feminist Media Studies*, 22 (8): 1972-89.
Hernandez, K. and Roberts, T. (2018), *Leaving No One Behind in a Digital World. K4D Emerging Issues Report*, Brighton, UK: Institute of Development Studies.
Hill-Collins, P. (1990), *Black Feminist Thought: Knowledge, Consciousness, and the Politics of Empowerment*, Boston: Unwin Hyman.
hooks, b. (2000), *Feminist Theory: From Margin to Centre*, London: Pluto Press.
Horn, J. (2025), *African Feminist Praxis*, London: Sage.
Iyer, N. (2022), 'Alternate Realities, Alternate Internets: African Feminist Research for a Feminist Internet', in A. Powell, A. Flynn, and L. Sugiura (eds), *The Palgrave Handbook of Gendered Violence and Technology*, Cham: Palgrave Macmillan.
Keller, J. (2018) 'Crop Tops and Solidarity Selfies: The Disruptive Politics of Girls' Hashtag Feminism', in M. Kearney, and M. Blue, (eds), *Mediated Girlhoods*, 157-73, New York: Peter Lang.
Kember, S. (2002), 'Reinventing Cyberfeminism: Cyberfeminism and the New Biology', *Economy and Society*, 31 (4): 626-41.
Kendall, M. (2020), *Hood Feminism: Notes from the Women White Feminists Forgot*, London: Bloomsbury Publishing.
Kennedy, B. (2000), 'Cyberfeminism: Introduction', in D. Bell, and B. Kennedy (eds), *The Cybercultures Reader*, London: Routledge, 283-90.
Kibona Clark, M. and Mohammed, W. F. (2023), *African Women in Digital Spaces*, Dar es Salaam: Mkuki na Nyota.
Loza, S. (2014), Hashtag Feminism,# SolidarityIsForWhiteWomen, and the Other# FemFuture. Available online at https://scholarsbank.uoregon.edu/xmlui/handle/1794/26991 Accessed 4/0/2024.
Luxembourg, R. (1908), *Reform or Revolution*, London: Militant Publications, https://www.marxists.org/archive/luxemburg/1900/reform-revolution/index.htm.
MacKenzie, D. and Wajcman, J. (1985), *The Social Shaping of Technology: How the Refrigerator Got its Hum*, Milton Keynes Philadelphia: Open University Press, ISBN 9780335150267.
Mama, A. (2002), *Beyond the Masks: Race, Gender and Subjectivity*. London: Routledge.
McLean, N. and Mugo, T. K. (2015), 'The Digital Age: A Feminist Future for the Queer African Woman', *IDS Bulletin*, 46 (4), https://bulletin.ids.ac.uk/index.php/idsbo/article/view/92.
Mejias, U. A. and Couldry, N. (2024), *Data Grab: The New Colonialism of Big Tech and How to Fight Back*, London: Penguin.
Mekgwe, P. (2006), 'Theorizing African Feminism(s): The "Colonial" Question', *QUEST: An African Journal of Philosophy*, 20 (1-2): 11-22.
Mendes, K., Ringrose, J. and Keller, J. (2019), *Digital Feminist Action: Girls and Women Fight Back Against Rape Culture*, New York: Oxford University Press.

Mhiripiri, N. and Moyo, B. (2016), 'A Resilient Unwanted Civil Society: The Gays and Lesbians of Zimbabwe Use of Facebook as Alternative Public Sphere in a Dominant Homophobic Society', in B. Mutsvairo (2018), *Digital Activism in the Social Media Era*, Newcastle: Palgrave Macmillan.

Mohanty, C. (1984), 'Under Western Eyes: Feminist Scholarship and Colonial Discourses', *Boundary* 2, 12 (3): 333–58.

Mohanty, C. and Torres, L. (eds) (1991), *Third World Women and the Politics of Feminism* (vol. 632), Bloomington, IN: Indiana University Press.

Molyneux, M. (1985), 'Mobilization without Emancipation? Women's Interests, the State, and Revolution in Nicaragua', *Feminist Studies*, 11 (2): 227–54.

Mpofu, S. (2017), 'Re-imagining Emancipatory Politics Online: A Critical Political Economy Approach of Women's Activist Media in Zimbabwe', *Communicatio*, 14 (3-4): 1–18.

Munro, E. (2013), 'Feminism: A Fourth Wave?', *Political Insight*, 4 (2), https://www.tandfonline.com/doi/full/10.1080/14680777.2015.1093070.

Mutula, S. M. (2008), 'Digital Divide and Economic Development: Case Study of Sub-Saharan Africa', *The Electronic Library*, 26 (4): 468–89.

Njoroge, D. (2016), 'Broken Silence: #Bringbackourgirls and the Feminism Discourse in Nigeria', in B. Mutsvario (ed.), *Digital Activism in the Social Media Era*. New York: Springer, 311–25.

Norman, D. (1988), *The Design of Everyday Things*, New York, NY: Basic Books.

Nyabola, N. (2018), *Digital Democracy, Analogue Politics*, London: Zed Books.

Nyamu-Musembi, C. (2005), 'An Actor-oriented Approach to Rights in Development', *IDS Bulletin*, 36 (1): 41–51.

Nzomo, M. (2003), 'Women and Democratisation Struggles in Africa: What Relevance to Postmodernist Discourse?', in M. Marchand and J. Parpart (eds), *Feminism/Postmodernism/Development*, London: Routledge, 145–55.

Ogbonna, E. (2018), *Digital Citizenship in Africa's Fractured Social Order*, London: Routledge.

Paasonen S. (2011), 'Revisiting Cyberfeminism', *Communications*, 36 (3): 335–52.

Roberts, P. (1983), 'Feminism in Africa: Feminism and Africa', *Review of African Political Economy*, 10 (27-28): 175–84.

Roberts, T. (2015), Critical Agency in ICT4D (Unpublished PhD), Royal Holloway University of London.

Roberts, T. (2016), 'Critical Intent & Interests: A Typology of ICT4D Initiatives', in *Proceedings of the 13th International Conference on the Social Implications of Computers in Developing Countries*, 20-22 May 2015, Negombo, Sri Lanka.

Roberts, T. (2021), 'Digital Affordances in Participatory Research Methods', in D. Burns, J. Howard and S. Ospina (eds), *The SAGE Handbook of Participatory Research and Inquiry*, London: SAGE.

Roberts, T. and Bosch, T. (2023), *Digital Citizenship in Africa: Technologies of Agency and Repression*, London: Bloomsbury Academic.

Rottenberg, C. (2019), '#MeToo and the Prospects of Political Change', *Soundings*, 71 (Spring): 40–9.

Salami, M. (2014), 'The Coming of (Digital) Age: How African Feminists are Using the Internet to Change Women's Lives', *GenderIT.org*, https://genderit.org/articles/coming-digital-age-how-african-feminists-are-using-internet-change-womens-lives.

Sobande, F. (2020), 'Why the Digital Lives of Black Women in Britain?', Springer online, https://link.springer.com/chapter/10.1007/978-3-030-46679-4_1.

Trott, V. (2021), 'Networked Feminism: Counterpublics and the Intersectional Issues of #MeToo', *Feminist Media Studies*, 21 (7): 1125–42.
Tufekci, Z. (2017), *Twitter and Tear Gas*, New Haven: Yale University Press.
Wajcman, J. (1991), *Feminism Confronts Technology*, Cambridge: Polity Press.
Wajcman, J. (2004), *Techno-Feminism*, Cambridge: Polity Press.
Young, K. (1993), *Planning Development with Women: Making a World of Difference*, London: Macmillan.
Zarkov, D. and Davis, K. (2018), 'Ambiguities and Dilemmas Around #MeToo: #ForHow Long and #WhereTo?', *European Journal of Women's Studies*, 25 (1): 3–9.

Chapter 2

QUEER FEMINIST DIGITAL CITIZENSHIP ON NIGERIA'S X: THE CASE OF #UJUANYA

Ochega Ataguba

This chapter examines how hashtags are resignified to enact political subjectivities and create spaces for the performance of queer feminist digital citizenship in Nigeria's X (Isin and Ruppert, 2020, Yue and Lim, 2022).[1] It engages with the discourses that emerged from the hashtag #UjuAnya, in response to a Twitter storm that erupted from the scathing tweet by Uju Anya, an unabashed lesbian Black, Nigerian woman and LGBTQ advocate. In that tweet, @UjuAnya wished the former Queen 'excruciating death', expressing disdain over her in the British Empire's exploitation and handling of the violence that led to civilian casualties during Nigeria's Biafra War.

The tweet racked up countless likes and shares, prompting a hailstorm of positive and negative reactions. Hours later, X deleted the tweet, which was perceived as highly offensive. Anya was hashtagified, and the hashtag spilt over into feminist and queer discourses, reigniting the debate about the exploitations of British colonialism and the misrecognition of queer women in Nigeria and their need to reclaim agency and positive identity (Olofintuade, 2017). At the same time, hypermasculine and anti-homosexuality campaigners decrying the erosion of African heteronormative cultures weaponized the hashtag, spreading a torrent of racialized and gendered identity-based attacks (Jane, 2014; Onanuga, 2021).

This incident offers an interesting case study of how feminist subjectivities are articulated on X by drawing upon the 171 tweets associated with #UjuAnya to provide insights into the reactions and responses the event triggered. The chapter shows how the very appearance of a queer woman is provocative in male-dominated, conservative social contexts, such as Nigeria, as evidenced by the visceral responses and negative reactions they receive. Nigeria's mainstream media privilege heteropatriarchal narratives and are invariably intolerant of non-heterosexual expressions (Ayodele Onanuga, 2020). As scholarly attention continues to cluster around X and its influence on civic engagement and social justice activism in Nigeria (Obia, 2020; Uwazuruike, 2020), questions have been asked about its impact on gender discourses and queer politics, visibility and activism, especially within the social subtext of heteropatriarchy (Strand, 2019; Onanuga, 2021).

Prior research has demonstrated the socially impactful use of feminist hashtags in forming alternate publics and countering hegemonic discourse around gender identity and sexuality (Drüeke and Zobl, 2018; Jackson, Bailey and Welles, 2020).

However, there is a tension between the liberatory promise of social media and the radical vulnerability it portends alongside the risk of harm and misappropriation (Johns and McCosker, 2015). Hence, scholars caution that research exploring feminist hashtags must be cognizant of the ways feminist activism is both enabled and constrained in digital spaces (Linabary, Corple and Cooky, 2019). Others have cautioned against a wholesale understanding of X as a digital public sphere without paying attention to how such a sphere can structure some discourses and identities to exist visibly and others to be backgrounded (Maragh-Lloyd, 2020). Hence, the chapter investigates how X's counter-narrative affordances enhance or complicate preexisting feminist practices and the vulnerabilities attached to publicly performing queer identity in the Nigerian context – a culture that strongly ties ideas of women's citizenship to propriety and religiosity and depoliticizes queerness (Akpojivi, 2019; Obadare, 2015). More specifically, it examines how digital subjects reconfigured the hashtag #UjuAnya to claim rights to sexual citizenship and queer visibility and to make claims as queer digital feminist citizens in this profoundly religious, hetero-heteropatriarchal social context (Yue and Lim, 2022; Isin and Ruppert, 2020). The chapter will demonstrate ways of understanding queer feminist subjectivities and Nigerianness in a digital environment marked by intense political debates over which bodies are worthy of making rights claims and which digital acts are considered dangerous transgressions that need policing and discipline.

The chapter addresses these questions

The chapter addresses the following questions:

- How do queer Nigerian feminists perform digital citizenship on X?
- How was #UjuAnya exploited for queer visibility and self-narrative gender affirmation?

The chapter critically engages with intersectionality (Crenshaw, 1991) and (feminist) digital citizenship (Isin and Ruppert, 2020) to untangle these questions. It expands upon notions that the digital space is gendered (Megarry, 2018; Sobieraj, 2020) and considers how X, as an instigative platform amenable to intersectionality (Zimmerman, 2017), offers opportunities for queer feminists to make rights claims against gender injustice and push back on their misrecognition (Zimmerman, 2017; Onanuga, 2021). X is a site of activism where queer Nigerians can contest the invisibility Nigerian society forces upon them (Ayodele Onanuga, 2020). While there have been recent works on hashtag feminism generally in the Nigerian context (Faniyi, 2023; Dosekun, 2022; Olofintuade, 2017; Faniyi, Nduka-Nwosu and Gajjala, 2023), scholarship on digital queerness is still nascent. Much existing research focusing on LGBTQ activism has concentrated more generally on the activism of gay and transgender individuals and its implication for queer representation (Nwabunnia, 2021; Ayodele Onanuga, 2020; Onanuga, 2021). However, research focusing specifically on queer Nigerian women on X, particularly lesbian women, has been relatively unexplored. This under-representation is unsurprising given the overall invisibilization of lesbian women in LGBTQ spaces and within research in various social contexts that results from their perceived socially subordinate role as women (Hildebrandt and Chua, 2017).

The chapter aims to fill this gap and extend the work of scholars who have already begun the process by drawing on a feminist approach to digital citizenship and intersectionality by positioning queer visibility as the ideal as part of an egalitarian society. The chapter argues that there are limitations to how far X's affordances can challenge misogyny, homophobia and other existing social structural barriers; however, tactics of resilience work as agentic strategies to enable queer feminists to enact their political subjectivities and make claims against gender discrimination and social injustice concerns that are at the heart of the feminist agenda.

Apart from this introductory part, the rest of this chapter is divided into four sections. The next section elaborates on the overall socio-political context of Nigeria in which the struggle over queer identities plays out. It also reviews the existing criminalizing law on same-sex relations in Nigeria before looking at the country's digital terrain, the use of X, and the case study that produced the hashtag under study. Next, it reviews existing literature on the trajectory of the feminism wave, digital feminism and queerness while offering a brief overview of feminism in Nigeria. The next section explores the theories of digital citizenship and intersectionality. However, this part does not attempt an exhaustive analysis of these theories. Instead, it identifies their relevance to the chapter, particularly as a precursor to analysis that will be discussed in the findings. The final section presents the methodology, findings and analyses and the emerging themes from the dataset before concluding the chapter.

Nigeria's political socio-cultural space, religion and sexual politics

Nigeria is a complex multi-ethnic and multi-religious country in West Africa whose northern and southern protectorates were amalgamated into a unified country in 1914 by the British colonial authorities (Hayward, 2019). It is also one of the world's most religious and conservative societies (Sunday, 2020). The northern part of Nigeria is predominantly Muslim, with twelve states operating under Sharia law. Muslims comprise about 50 per cent of the population, with Christians and those engaging in traditional religions accounting for 40 per cent and 10 per cent, respectively (Brinkel and Ait-Hida, 2012). Hence, Islam and Christianity comprise a substantial population of Nigeria and form a significant part of the country's cultural identity and ideologies, including the deeply held belief that homosexuality offends public morals. Patrick (2016) showed that, despite contestations that homosexuality is a Western import and thus un-African, homosexuality existed in precolonial African societies. The spread of Abrahamic religions, particularly fundamentalist Christianity by the British, meant that Nigerian society, like most parts of Africa, lost its previous cultural attitude towards sexual orientation, which was more accommodating towards non-normative sexualities, since it was forced to adopt and comply with the dominant binary sex/gender model of Eurocentric modernity (Buckle, 2020).

Aside from the impact of these religions in reinforcing heterosexism, the colonial state established the Queensland Criminal Code, which was first introduced to the Protectorate of Northern Nigeria in 1904, and later made applicable in the Southern region and Lagos colony in 1916 by Frederick Lugard, the colonial governor (Alkali et al., 2014). This legislation negatively impacted African sexual politics, and it was women's sexual autonomy that was most severely under threat (Tamale, 2011). These laws contained the penal code, part of which were anti-sodomy laws that criminalized and punished homosexuality under the provisions for unnatural offences (Watch, 2013). Through these processes, the colonial state effectively silenced precolonial tolerance and acceptance of diversity in sexual relationships in Nigeria (Sogunro, 2022). In this way, the struggle to end colonial rule was inextricably bound up with the battle against the trauma of state-sanctioned imperialist projects that expunged the indigenousness of non-normative identities in Nigeria reducing them to a homogenous right way of being citizens.

Meanwhile, colonial rule entrenched stratifications along the lines of class, gender, regional differences, ethnicity and the larger socio-political and economic conditioning that undermined basic citizenship rights (Diamond, 1988). When the country finally achieved political independence in October 1960, it allowed for further demands for greater self-determination and citizenship inclusion by various ethnic groups and regions. Owing to Nigeria's heterogeneity, with nearly 220 million people and 250 ethnic groups speaking over 350 languages, the fear of political domination by other ethnic groups became prevalent (Diamond, 1988; Falola and Heaton, 2008). It led to the violent Nigerian Civil War, also called the Biafra War, which broke out in Nigeria from 1967 to 1970 when the Igbos in the southeast decided to secede. Given the country's historical precedent in the context of British colonialism, British authorities fuelled the violence by supplying arms and materiel to the Nigerian military forces at the expense of Nigerian civilians (Baxter, 2015). The war resulted in a significant loss of lives, and estimates suggest that over three million Igbos died in combat, including children, who died mostly from starvation and disease (Baxter, 2015). The Igbos, the main ethnic group in the secessionist state of Biafra, claim that they are being covertly alienated from political fields (for instance, proscribed from running for the presidency), stripping them of their political power, agency and status as legitimate members of society (Baxter, 2015; Falola and Heaton, 2008).

The lingering memory of starvation as a weapon of war and the collective trauma of the period remain in the consciousness of the Igbos. This shared memory also inspired a sense of victimhood that has been transmitted to subsequent generations who continue to battle their identity, with memories of the oppression as unbearable scars inflicted on their psyche. Given this history, it is unsurprising that Uju Anya, an Igbo woman from the southeast region, expressed pent-up rage in her tweet about the former British Monarch, demonstrating how this trauma from the Nigerian Civil War memories perhaps reached a breaking point during these moments of outspokenness and reaction through the tweet. It shows how the devastation of the civil war events continue to be a source of pain rooted deeply in postcolonial subjects' memories and how traumatic memories are successively passed to generations who did not witness the traumatic event yet participate in the trauma of the memories that are intergenerationally handed down (Caruth, 2016). She was vulnerable to the harsh realities she was faced with as a queer Black Nigerian woman in the diaspora. Yet she was also sharp, tough and empowered in ways that few upper-middle-class Nigerian women in these precarious positions were especially after having publicly disclosed her sexual identity perceived to be in opposition to the gender binary allowed by patriarchy.

Nigeria's Same-Sex Marriage (Prohibition) Act

Despite increased global conversations around gender equality and the high visibility of feminist discourses and queer movements (Gill and Orgad, 2018),[2] Nigeria has been slow to adopt gender-progressive practices (Akinbobola, 2019; Olofintuade, 2017). Hence, debates around queer Nigerian women have been almost non-existent, a consequence of viewing non-heterosexuality as unnatural in addition to treating queerness as deviant (Obadare, 2015; Alichie, 2023). Given these negative attitudes towards sexual minorities in the country, there is a popular quip that despite ethnic cleavages and the religious intolerance that divides the country, a shared hatred for members of the LGBTQ community is a common thread that ultimately unites Nigerians (Job, 2020; Epprecht and Egya, 2011). To this effect, the Same-Sex Marriage (Prohibition) Act 2013 (SSMPA), signed into law in 2014 by President Goodluck, was backed by strong public support (Onanuga, 2021). The law criminalizes LGBTQ persons and imposes a fourteen-year jail term on anyone who engages in LGBTQ advocacy, participates in gay clubs or supports the activities of such organizations (Sogunro, 2022).

Violence and extrajudicial killings perpetrated against LGBTQ Nigerians have spiked since 2014, after the law was enacted (Alichie, 2023). Abusers who carry out these crimes evade prosecution, and the Nigerian state has so far proven to be complicit in the egregious act of homophobic violence (Alichie, 2023; Osinubi, 2016). Political rhetoric has also become more severe in tone. For example, in 2022, a video of Datti Baba Ahmed, a member of the House of Representatives, declaring that gay and gender-fluid individuals in Nigeria should be killed resurfaced on X (Lardy, 2022). It led to intense debates surrounding ideas of conservative values and the broader political and social cultural conditioning of stigmatization and violent exclusions from citizenship, which continue to generate controversy. More recently, a Nigerian Sharia court in Bauchi imposed a death sentence for homosexuality, sentencing three men to death by stoning in July 2022 (Hazzad, 2022). Lawyers did not represent them, and this absence of lawyers breaches their citizenship and constitutional rights (Hazzad, 2022).

Thus, queer individuals in Nigeria, like elsewhere, experience higher incarceration, murder, greater poverty, displacement, unequal access to the public sphere and disproportionate health concerns (Armisen, 2016). Scholars have used the concept of political homophobia to describe the type of state-sanctioned violence and strategic use of anti-homosexuality rhetoric and anti-LGBTQ legislation to incite violence against sexual minorities and to wield political influence (Bosia and Weiss, 2013). Sogunro (2022) shows how this is often the case in societies facing socio-economic pressure and legitimacy crises, demonstrating how Nigeria's state actors utilize anti-gay sentiment as the default straw man to deflect public attention from other pressing issues on the ground. The United Nations General Assembly Declarations on Sexual Orientation and Gender Identity have expressed deep concern about the violations of human rights that undermine the dignity of those subjected to homophobic and misogynistic abuses and the need to exercise fundamental freedoms based on sexual orientation. It draws largely on the Yogyakarta Principles, which emphasizes the freedom to express oneself, one's identity and one's sexuality without the state's interference (O'Flaherty and Fisher, 2008). However, the UN's declaration was not binding since, just like the Millennium Development Goals, it only represented the ideals, aspirations, goals and targets that may or may not be enforced. Hence, the UN declaration received antagonism and cynicism from state and non-state actors (Akanji, Epprecht and Nyeck, 2013).

Quite often, Nigeria's civil society and women's rights organizations as a collective have been somewhat indifferent and slow to respond to or challenge these repudiations of queer practices and the SSMPA in the same ways they respond to other social justice or feminist issues that concern cisgender women (Nwabunnia, 2021; Sogunro, 2018). Sogunro (2018) recognized this and queried how their indifference to the discrimination that queer people experience in Nigeria stands in marked contrast to their outspoken approach to laws that propose regulation of social media (Oladapo and Ojebuyi, 2017) or police brutality and extortion as it were (Uwalaka, 2022). These indifferent attitudes and strategic silences stem from a lack of intersectional thinking in the current feminist discourse that is largely shaped by deeper systemic issues of political homophobia, revealing the barriers still in place to trans inclusion. Furthermore, it demonstrates how appeals to solidarity in social justice activism based on one identity often privilege the dominant majority group at the expense of the minority, thus producing exclusionary solidarity rather than allyship with multiply marginalized others (Ferree and Roth, 1998). Studies on social activism have highlighted how conflicting identities have historically relegated Black lesbian women and trans people, as well as people with disabilities, to the margins (Hooks, 1981; Yuval-Davis, 2006).

For example, in the wake of #EndSARS in Nigeria, when LGBTQ protesters raised placards saying #QueerLivesMatter and #EndSARS alongside the rainbow flag to address the SARS unit's homophobic violence, they were attacked by some members of civil society and accused of hijacking the #EndSARS movement (Job, 2020). Criticisms were directed at feminist groups such as the Feminist Coalition, which supported members of LGBTQ, accusing it of co-opting #EndSARS into queer activism, perceived as a 'demonic agenda' (Owoh, 2020). Although LGBTQ organisations, such as Safe House, provided support and safe housing for displaced queer individuals during the protest, that contentious episode heightened homophobic violence and conflict between some queer feminists and cisgender feminists who privileged cisgender feminism while framing queer feminism as improper (Nwabunnia, 2021). This privileging of heterosexual feminism over queer feminism results from the ideological tensions between cisgender feminists and queer feminists in Nigeria (Toyo and Adegbeye, 2021). The rise of queer activism and its push for lesbians and trans people, especially those who are biologically male, to be included in feminist organizing with cisgender women is a highly contentious issue that has created fractures, tensions and resistance within these feminist spaces in Nigeria. As a result, queer women are often in a conundrum and forced to choose or separate their 'womanness' from their queerness as a strategic approach needed to fit in or survive within the public space. It is instructive for feminists to revisit their understanding of intersectionality approaches (Crenshaw, 1991) and focus on marginalized groups who often feel underrepresented, excluded and undervalued in feminist movements (Collins, 2015). Despite these precarious circumstances of ostracization, digital platforms can provide alternative spaces for feminist action where pro-LGBTQ narratives can de-incentivize the invisibility of physical spaces (Onanuga, 2021; Yue and Lim, 2022).

Nigeria's social media landscape and the use of X

Since the popularization of the internet in Nigeria in 1996 and the deregulation of the country's communication industry in 2001 (Nkordeh, Bob-Manuel and Olowononi, 2017; Adomi, 2005), growing internet accessibility in the country has provided spaces for all individuals to interact and engage in various discourses (Novak, 2016; Obadare, 2021). In the first quarter of 2024, there were 103 million internet users in Nigeria, and internet penetration stood at 45.5 per cent (Kemp, 2024). There were 36.75 million social media users in 2024, of which 5.75 million Twitter /X users are in Nigeria. This is a relatively small percentage and is not representative of the country, but WhatsApp and Facebook have remained the most-used platforms in Nigeria (Odili, 2021). Although X does not have the most active user base in Nigeria, it is the most popular for digital activism (Obia, 2020) including feminist and social justice activism, making it the most suitable platform to carry out this study.

Fourth-wave feminism and digital queering

Much of the contemporary literature on digital feminist and gender and sexuality activism uses the fourth wave (Munro, 2013) to name and frame the opportunities and affordances digital spaces offer for feminists to circulate gender discourses and challenge the political and social order (Cochrane, 2013; Chamberlain, 2016; Clode, 2018; Baumgardner, 2011; Megarry, 2020a). The wave model signifies the generational accounts of the women's liberation movement in the United States, with the first wave of feminist movements beginning in the late nineteenth century with women's struggle for suffrage rights (Chamberlain, 2016). Second-wave feminists were concerned with issues such as their reproductive rights. However, the second wave collapsed in the 1980s due to ideological schisms, evident in its neglect of Black women's peculiar struggles and almost exclusive focus on middle-class white women (Vickery, 2018). Hence, it failed to acknowledge that Black women's experiences differed from those of white feminism since many Black African women were affected not only by sexism but also by the oppressions of racism, sexism, classism and colonialism (Lorde, 2012; Collins, 2022; hooks, 1981; Tamale, 2020). Second-wave feminism was also hostile to queer practices (Hines, 2005).

The queer-inspired third wave emerged from these contestations, challenging the homogenization of third-world women and reconceptualizing binary frames of sex and gender identities (Mohanty, 1988; Davies, 2018). Although many issues of interest for the fourth wave of feminism are not so distinct from issues tackled in the second and third, what characterizes this fourth wave of feminist mobilization is the global use of the internet, especially social media, to connect, learn and collaborate (Cochrane, 2013; Loza, 2014). Hence, Zimmerman (2017) argues that this new iteration of feminism distances itself from its predecessors with an emphasis 'on fighting racist and sexist oppressions through an intersectional feminist lens that considers social networks an indispensable and essential tool and strongly resists separating the offline from the online' (Zimmerman, 2017: 55).

However, some scholars are wary of oversimplifying feminism into four waves since it tends to flatten the multiple trajectories of women's rights activism, overlooking feminist struggles in other parts of the world, especially in the Global South (Mohanty, 1988). For example, the metaphor hardly reflects the progression of the Nigerian feminist movement. Regardless, the Nigerian feminist movement is not definitive since the contributions of feminists in Nigeria have been a mix of continental and diaspora global perspectives rather than a uniquely Nigerian one (Salami, 2018). Still, the largest Nigerian women's movement was the National Council of Women's Societies (NCW), founded in 1958, coinciding with the second wave of feminism in the West. Yet, the NCW was not effective because it was complicit with patriarchy and was only interested in the upliftment of women in spheres of motherhood, family and contributions to nation-building (Madunagu, 2008).

The movement was also largely silent around sexual rights and was deeply homophobic. In 1983, another organization, Women in Nigeria, emerged, and many believed it understood the feminist agenda. It was open to different ideological views and engaged in advocacy and activism to transform the conditions under which women and underprivileged classes in Nigeria lived. Building from Women in Nigeria's work, the Nigerian Feminist Forum emerged in 2006 with an undeniably egalitarian outlook and a goal to resist patriarchy and uplift marginalized identities (Madunagu, 2008). The activities of the Nigerian Feminist Forum also coincided with the proliferation of social media in Nigeria, which gave rise to a sudden shift in the tone of the conversation around feminism and gender discourse in the country (Osibu, 2022).

Indeed, the impact of social media in advancing feminist discourses in African societies cannot be overemphasized (Nyabola, 2018). This growing use of social media heralded the concept of hashtag feminism (Loza, 2014; Jackson, Bailey and Foucault Welles, 2018), capturing the explosion of feminist hashtags in specific moments of resistance. Prominent hashtag movements in Nigeria, such as #BringBackOurGirls and #BeingFemaleInNigeria (Akinbobola, 2020), were amplified transnationally, exposing misogynistic violence, oppression of women, molestation and terror with a worldwide audience bearing witness (Berents, 2016). These online activities led to material effects, translating into human action in offline spaces despite the perceived immateriality of digital spaces (Njoroge, 2016). Bailey (2021) and others have raised concerns about how women outside the dominant gender binary are excluded from feminist movements. For example, Nanditha (2022) shows how non-inclusivity was evidenced in the sparsity of personal narratives in India's #MeToo and the harassment of trans women. These issues have provided opportunities to challenge and critique the internal dynamics of feminist movements, enabling new kinds of intersectional conversations (Linabary, Corple and Cooky, 2019; Baer, 2018). From this, it seems crucial to better understand the implication and usefulness of social media for non-binary individuals who suffer hostile forms of misrecognition and often lack the opportunity for offline support (McCosker, Vivienne and Johns, 2016).

For example, sexual minorities in many African societies who bear the brunt of discriminatory attacks and anti-gay violence rely on digital spaces such as X as sites for liberatory mobilization (Strand, 2019; Fotopoulou, 2012; Mwangi, 2014). Hence, X and its affordances of hashtag activism open spaces for self-assertion for sexual minorities to contest the invisibility society forces upon them and how they may build affinity with the collective (Onanuga, 2022). Those suffering rejection and ostracization because of their gender identity or sexuality feel the need to reconstruct their experiences in a manner that gives them voice, agency and positive identity (Nartey, 2022: 2). Thus, it suggested that sexual minorities must render themselves visible because being visible is associated with reclaiming agency and owning one's true identity and sexuality (Hanckel et al., 2019).

Some scholars, such as Clark Parsons (2022), have used the phrasing 'politics of visibility' to describe how feminist hashtags enhance visibility in a way that powerfully shifts how activists represent, react and respond to social injustice through public performances (Clark-Parsons, 2022). In the sense of coming out of the closet, visibility is also conceptualized as creating space for alleviating feelings of shame, anger or loneliness. It is also required to advocate for rights and to forge coalitions and networks (Pindi, 2020). While X affordances can serve as a tool for public self-making and representation for queer individuals, feminist scholars have also emphasized that increased online visibility increases the risk of further stigmatization and vulnerability (McCosker, Vivienne and Johns, 2016). Meanwhile, Fotopoulou (2017) notes that vulnerability can be a productive concept with political potential when using digital tools. Hence, it should not always be seen as victimization or as passivity but as an important arsenal for feminist and queer activism because it offers great opportunities to articulate powerful responses and reactions that can build resilience in the face of oppression (Fotopoulou, 2017).

Conversely, in social contexts where the sexual minority community has limited opportunities to self-represent in a non-discriminatory manner, societal discourse actively contributes towards the community's vulnerability. For example, while X is an inherently public, interaction-driven platform, sexual minorities in Uganda do not take advantage of X's inbuilt affordances for visibility (Strand, 2019). Rather, X is approached as an intra-community platform; in other words, it is used for private, siloed community network interactions to avoid homophobic backlash. In addition to the prevailing onslaught of homophobia, violent misogyny is deployed to compound online abuse of non-binary women, making feminist work all the more precarious (Jane, 2014). Megarry (2018) has drawn attention to how white heterosexual men govern social media spaces. In this sense, the tools that facilitate digital feminist action also tend to undermine it because digital spaces privilege masculine-oriented, heteronormative whites and distort all other perspectives by default (Leurs and Ponzanesi, 2011). Chemaly's (2018) work describes the rubber stamp of hierarchies placed by platforms like X that render feminist activists' work and that of marginal groups invisible. Because of this, not all women are listened to equally on social media, and those who command attention via sharing and liking functionalities are likely to be women who present as middle class, able-bodied and heterosexual (Megarry, 2020b). Despite these accumulating challenges and failings of social media platforms to live up to their democratizing expectations, social media still seems to have considerable promise in enabling digital feminist action. It has also enabled sexual minorities to develop a sense of community and belonging online, as it provides possibilities for LGBTQ communities to reach like-minded diverse public and bring to the fore multiple intersections that shape their identities (Ayodele Onanuga, 2020). In sub-Saharan African societies, such as Nigeria or Uganda, for instance, where attempts to use physical spaces as platforms for advocacy have been met with fierce and even violent resistance in the past decades (Nyanzi, 2014), the barriers to performing feminist citizenship are still lower in the digital public sphere than in offline civic spaces (Yeku, 2020; Dambo et al., 2022; Job, 2020). Still, this does not diminish the crosshairs of harm and injustice perpetuated against sexual minorities and Black women who face violent opposition in public discourse, where people attack their identities rather than take issues with their ideas (Sobieraj, 2020).

Connective frameworks: Intersectionality and digital citizenship

Intersectionality is an analytical strategy emphasizing how race, class, gender and sexual orientation are mutually constitutive social identity categories that are not conceptually distinct but cross-cutting and entangled within multiple discrimination scales (Collins, 2015). The concept was introduced by Kimberlé Crenshaw (1989) to address the failure of the US justice system to recognize how these multiple social identity categories, compared to just one dimension of identity, combine and compound Black women's experience of oppression, discrimination and domination (Crenshaw, 1989; Yuval-Davis, 2006).

In this chapter, intersectionality is explored through the prism of digital citizenship as claims-making and used to emphasize how queer feminists who are multiply marginalized make rights claims against oppressive heteropatriarchal discourses that continue to circulate on X. In the digital spaces, barriers to entry have been made lower, but platforms prioritize whose voices are heard and reify power hierarchies (Murthy, 2013). Thus, there is a need to recognize and address how the social hierarchies of race, class and ethnicity that are reflected online are reformed and resisted (Sobieraj, 2020), through an intersectional approach to guide the analysis of the workings of X for queer feminist digital citizenship performances (Dosekun, 2022).

Central to the notion of digital citizenship is the process of claiming rights and the right of citizens to give an account of themselves despite discriminatory frameworks of conservative sentiments that deny voice and agency (Couldry, 2010). Here, rights claim-making is a 'perspectival process' that entails expressing civil disobedience, transgressions and subversion while contesting and negotiating polemic narratives to destabilize hegemonic ideologies (Zivi, 2012: 51 & 67).

Though the basic understanding of digital citizenship is the ability to access and use digital media technologies and devices to participate in society (Mossberger, Tolbert and McNeal, 2007), this chapter goes beyond this understanding to focus on the enactment of political subjectivity in cyberspace and the rights claims queer feminists make through forms of transgressions, assertions or political speech (Isin and Ruppert, 2020). Isin and Ruppert (2020) persuasively demonstrate that individuals and groups previously excluded from citizenship can assert themselves as claims-making subjects by acting up against repression, domination and injustices on the internet through the process of subjectivation wherein they appropriate digital media affordances to post, comment, reply, share, block and perform counter-narrative digital acts.

In this way, digital citizenship as claims-making highlights the potential for destabilization that online contextualization and digital activism provide. By bringing the concept of opening and closing, which highlights how X can undermine and enable participation and engagement on X to the centre of analysis, it interferes with the techno-determinist thinking implicit in hyperbolic assertions of social media's impact that imagine subjects as passive digital subjects. This way, digital citizenship reconnects to the broader political, economic and social realities that frame digital technology, such as the algorithmic protocols that govern social media use. Moreover, Isin and Ruppert (2020) contend that if we shift our focus from how we are being controlled to the complexities of acting as digitally enabled subjects, not in isolation but to the arrangements of which we are a part, we can identify ways of being subversive.

Intersectionality, on the other hand, is particularly relevant for analysis given that Nigerian women constitute a hugely diverse group, differentiated by sexual identities and ideologies and peculiar oppressions that result from ethnicity, religion, socio-economic class, self-repression, cultural and traditional subjugation, heteropatriarchy and poverty that intersect to include, exclude and undermine women (Pereira and Ibrahim, 2010; Faniyi, Nduka-Nwosu and Gajjala, 2023). As a composite of multiple subjectivities that emerge from different situations and contexts, how we conduct ourselves in cyberspace through digital acts is highly dependent on these overlapping identities, and allegiances are all too familiar, complicating the expression of full citizenship and the enactment of queer feminist digital citizenship (Isin and Ruppert, 2020). Ogundipe-Leslie's (1994) work on Statism (Social Transformation in Africa Including Women) considers intersectionality particularly useful in the Nigerian context because, if carried out effectively, the coalition forces imbued with an intersectional approach can foster the inclusion of multiply marginalized queer women within the coalition.

Research methods

This study aims at uncovering how the hashtag #UjuAnya was used to perform queer feminist digital citizenship. The study used a digital methods (Rogers, 2019) approach to qualitative research, utilizing Twitonomy to retrieve 721 tweets between 8 September and 30 October 2022.³ Tweets were collected using the hashtag. Heated conversations on X tended to have currency for a few days after the 8 September tweet. Hence, the researcher considered four more weeks a reasonable time frame to gather tweets that captured the essence of the conversations around #UjuAnya. After removing tweets that did not have interpretable content, the number of tweets analysed for this chapter was n =171, as illustrated in Figure 2.1. It was observed through profile information that the tweets were predominantly from Nigerian users (86.4 per cent), with the rest spread over the United States (4.5 per cent) and an unknown 9 per cent.

The data was analysed using thematic analysis (Braun and Clarke, 2021) to identify and report issues and patterns that emerged from the data. Thematic analysis is a realist method that helps to reflect the reality of X users and how they make meaning of their experiences. The six steps in Braun and Clarke's (2021) approach to reflexive thematic analysis were carried out. Eight codes were identified in response to RQ1 and RQ2. Then, coding categories were created and collapsed into three overlapping themes. These subcategories discussed in the next section were created from these themes. These include (1) Intersectionality

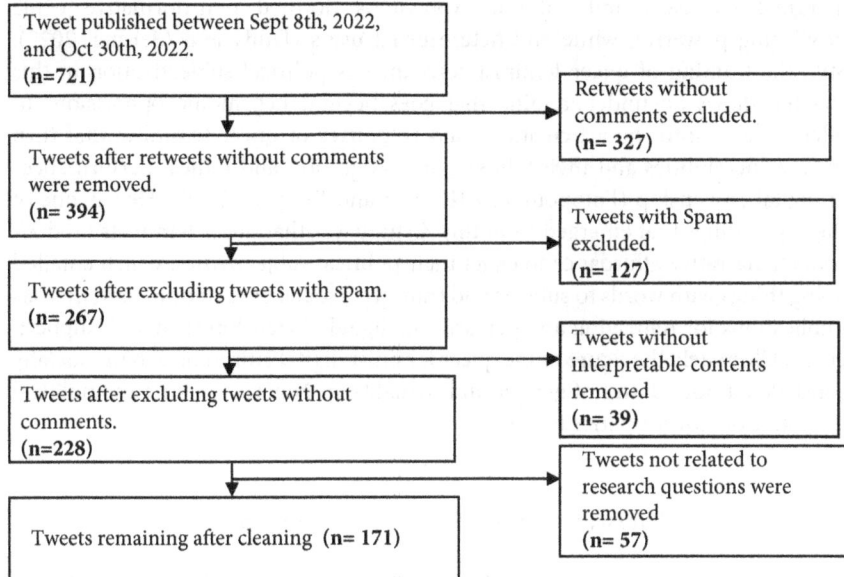

Figure 2.1 A visual breakdown of data screening and selection shows the tweets in the dataset for #UjuAnya

and the tactics of resilience, (2) Defiant visibility and solidarity building and (3) Homophobia and misogynistic violence.

Findings from X and discussion

Nigeria's X has emerged as a robust space where discourses on queerness, gender and sexuality are increasingly being challenged within the presumed normativity and social and political culture at odds with it. This dataset revealed the lesbophobic and misogynistic outrage that pervades this predominantly heteronormative social context but also illustrated how X serves as a space for performing queer feminist digital citizenship through the subversive action possibilities it offers. From the audacious, resilient use of the counter-narrative affordances of X, queer feminists made rights claims (Zivi, 2012) against gendered oppression and homophobic violence, and this gave new visibility and urgency to deeply rooted issues of gender injustices and homophobia in Nigeria. By bringing into focus Black African women within queer performances, #UjuAya provided space for non-binary Nigerians to gain a form of visibility.

However, the dataset also indicates that there is a tension between X's use in queer feminist work and the platform's complicity in promoting sexism, misogynistic and heteropatriarchal ideologies (Alichie, 2023). Also, the regimes of control and punishment that structure X's platform economy tend to disadvantage marginalized users and cultural expressions deemed non-normative while privileging powerful, white and heterosexual users (Duffy and Meisner, 2023). Still, the framing of queer feminist resistance as political subjectivation in this chapter allows an understanding that goes beyond hegemonic oppressions to offer insights into the articulations and responses of queer feminists that turn their vulnerabilities and invisibilities into recognition and radical performances of digital citizenship (Fotopoulou, 2017; Isin and Ruppert, 2020). Hence, one of the key findings that emerged from this X study was that queer feminists used X's counter-narrative affordance to enact their political subjectivities, which entailed doing things with words to subvert and transgress social expectations, and this has implications for reimaging and performing digital citizenship (Isin and Ruppert, 2015). Ultimately the figure of the queer feminist digital citizen on X blurs socially constructed boundaries of gender and sexuality in Nigeria, making the political discourse on queer politics fiercer.

Intersectionality and the tactics of resilience

In this analysis, many X users who participated in the hashtag engaged with the first tweet that triggered the maelstrom, in which @UjuAnya had criticized the British monarch, and @JeffBezos responded almost immediately, provoking X outrage as shown in these tweet examples:

> The tweet: *I heard the chief monarch of a thieving, raping, genocidal empire is finally dying. May her pain be excruciating. @UjuAya*
> Reply: *This is someone supposedly working to make the world better. I don't think so. Wow. @JeffBezos*

Amid the uproar and contention surrounding the perceived insensitiveness of the tweet, the confrontation between @JeffBezos and @UjuAnya drew attention to the power dynamics at play, considering how digital spaces as hypermasculine environments (Megarry, 2018) are complicit in gendered oppression, especially when directed at Black gender non-compliant women who speak out in public and transverse the expectation of propriety. It accrues from the exclusive heteronormative and male supremacist power structures that undergird heteropatriarchal societies. Foucault (1978) underscores the relationship between power and sexuality and argues that sexuality is an especially dense transfer point for relations of power. It becomes imminent when considering the regulation and limits of gender, sex and sexuality. In this instance, Bezos's hyper-rational tweet can be regarded as a loaded and gendered form of manipulation, a silencing strategy enacted against women to close opportunities for participation in the digital public sphere (Sobieraj, 2020). Explicitly rejecting the gendered surveillance discipline that came with the tweet, @UjuAnya replied to @JeffBezos:

Otoro gba gbue gi May everyone you and your merciless greed have harmed in this world remember you as fondly as I remember my colonizer. @UjuAnya

Not all users associated with the hashtag were feminists or used it to advance feminist views. Hence, it generated opinions and reactions that were liberatory and oppressive (Linabary, Corple and Cooky, 2019). Soon, the hashtag took on a life of its own through the emotional resonance established with other feminists, queer individuals and allies, and those who shared cultural understandings of the horrid experience of the Nigerian Civil War @UuAnya had tweeted about. It was unsurprising, given that @UjuAnya's huge following on X is predominantly Nigerian. Isin and Ruppert call attention to how 'hashtags are resignified to name and mobilize social movements' (Isin and Ruppert, 2020: 1), creating openings for multiple voices to participate in the conversation despite the tendency of detractors to transform the hashtag into an appendage of a hate campaign.

Initially, it was the feminist detractors who hashtagified @UjuAnya's name. Many tweets (*n* =39; 22.81 per cent) were critical of her choice of words in the tweet, which was perceived as 'insensitive', 'mean' and 'hostile'. In those tweets, she was denigrated and threatened for posting the tweet and was ordered to take it down. For example, a user tweeted the following:

When urged to delete the tweet and apologize, @UjuAnya tweeted a response in which she wrote:

If anyone expected me to express anything but disdain for the monarch who sponsored the genocide that massacred and displaced half of her family, they were wishing upon a star. @UjuAnya

Some tweets *(n = 17; 9.94 per cent)* presented arguments in support of @UjuAnya and her accusation against the British monarch. While many users abused, shamed and badgered her for the post, others defended it and supported her standpoint, as in the following tweet:

God save our King
@Godsaveourk

Delete your tweets about the queen or face the consequence. You have been warned , we know your face, name and exactly what you stand for. If the establishment doesn't ruin you. We will. you really have messed up now

5:57 AM · Sep 10, 2022

Figure 2.2 Threatening response to @UjuAnya's tweet

Wishing someone a bad death when they are dying is not racist, nor does it incite violence. It targets a public figure and a powerful one. It's not even false: that's an unkind, historically accurate statement of the British Empire and her formal role in it. @Ardo

Tweets like this one expressed that @UjuAnya's tweet did not violate community guidelines but was a harsh expression of her opinion. These tweets point towards what scholars (Sobieraj, 2020) argue about the freedom to talk without judgement and the need for an equitable and inclusive digital space where women and LGBTQ individuals can express themselves freely. Other tweets argued that it was a hostile, inciting tweet, and not only should it be taken down but Anya's account should be suspended. Twitter took down the tweet, claiming it violated the platform's rules. The removal of the post by Twitter illustrates how criticisms of powerful people can be eliminated from social media sites for political reasons, highlighting the power of digital platforms to reinforce biases and inequities (Biddie, 2022). While it can be argued that all women encounter forms of discrimination, the stakes are amplified for lesbian women in social contexts like Nigerian Twitter, where homosexuality is illegal (McCosker, Vivienne and Johns, 2016). A queer Nigerian woman like Anya is, therefore, not only challenging norms of gender and sexuality but opening up a locus where queer individuals can subvert prejudice and hegemonic orders.

Her refusal to take down the tweet and apologize in the face of intimidation and threats is also key. Such an act of resistance and resilience can also be seen as civil disobedience through the prism of Isin and Ruppert's (2020) work since it challenges the platform's authority. This digital act is also an oppositional response to prescribed norms of gender and identity where women are generally expected to be nice. The ability to cast aside demands of stereotypical feminine behaviour and act defiantly is a symbolic assertion of the transgression of social expectations. The findings that, despite their limitations, online spaces afford women and queer individuals the ability to transcend the limitation of offline spaces were confirmed (Sobieraj, 2020).

The British govt needs to apologise for its role in the destruction of Nigeria. Where is the lie? All I see is Twitter and white supremacy throwing tantrums because a black woman spoke! The rage is real, and the pain and loss are real. @Uzooma

Haba, She is uncouth! Rude as it is natural for Black women and a Nigerian, fa. Disgrace. @adugaba

No, this statement is out of step! Lesbian women that lack morals @Ezimbe

Reactions in the latter tweets hint at the perceived crossed boundary and violation of the place of women in society because speaking as a woman in public is perceived as inherently transgressive, as it steps outside the perceived gender boundaries that strive to limit and silence women. In this instance, the dimensions of Anya's social identities combine, subjecting her to multiple modes of oppression (Crenshaw, 1989). The visceral nature of the abuse and identity attacks directed at Anya because of her identity aligns with the conclusions of feminist scholars (Sobieraj, 2020; Loza, 2014) who argue that Black women who self-identify as feminists or are non-compliant with gender norms or challenge the status quo in male-dominated circles bear the brunt of digital hate threats and attacks.

Defiant visibility and solidarity building

The dataset also shows how Anya used her X account as a platform to challenge the exclusionary practices that diminish the humanity of queer individuals in Nigeria and to build solidarity with other members of the LGBTQ community in transgressive ways.

On her X account, @UjuAnya often used Igbo and Pidgin English to establish connections with her Nigerian followers and queer individuals who look up to her as an inspiration. She uses her visibility not just to engender queer discourse but also to encourage others who have experienced discrimination and oppression. In a popular retweeted message from Anya, which was retweeted 120 times, she offers a positive, hopeful message to the Nigerian LGBTQ community. She writes,

> Not everyone has the safety, security, or support to be out! So we see your struggles, and we love you! @UjuAnya

While posting these tweets, she shared a picture of herself and her female partner @ProfAlang wearing expensive-looking clothes decorated with a red heart emoji, hence the word 'we'. This tweet from Anya represents many of the tweets within the network where words of affirmation show solidarity and support among sexual minority users who come to her Twitter/X page for inspiration. The tweet was retweeted 27 times. Tweets such as this were accompanied by words of gratitude from her followers and other Twitter users who are members of the LGBTQ community.

Replies from numerous queer individuals, as well as heterosexual detractors, flooded the comment session. More support and advice chime in:

> For everyone who has endured bigotry just because of who they choose to love, I see you, and there is hope.

> Thank you for using your platform to create awareness about what we face here as queer Africans. You rock!
> @Amaka

> I just shared your handle with my partner. Your page here has been a balm when I felt isolated and alienated. I could never stand Twitter, but I am here daily, using my name because of you. That means a lot (heart emoji)
> @Winniw

> *I dare not post the real picture of myself and my partner here. I don't know what would happen. Thank you for being bold and brave! You guys are gorgeous. An inspiration to us all. @AppleTart16*

> *Oh please! All you women seeking victimhood carry your attention-grabbing out of feminism on Twitter. Go find work to do.*

Several cynical comments like this last one were blocked or counter-attacked with expletives and scornful replies. The way the comment attempted to silence the queer feminists is related to the ways the public sphere is characterized by hegemonic masculinity that dictates what types of normative displays of femininity are allowed and which ones are not. Unlike in societies such as Uganda (Strand, 2019), where the affordances of X are not consistently exploited for visibility owing to the fear of backlash and hostilities, in heteronormative environments in Nigeria, X seems to provide some wiggle room for queer individuals to boldly contest these invisibilities where sexual minorities seek to write themselves in places where they are not represented. Queer citizenship performances on X challenge not just norms of gender and sexuality but also open up a lotus for sexual minorities in Nigeria's digital public sphere. Through these processes, Twitter/X becomes a space where feminists online can transgress the institutionalized gender norms.

Anya's position as a Nigerian lesbian woman is a rarity in a conservative society that wants lesbians hidden and erased. Hence, it can be argued that her visibility and assertiveness were revolutionary. She used her platform to project her activism against social constructions that exclude and disregard LGBTQ people by creating a space where she frequently posts about sexual freedom and the need to dismantle patriarchal systems. Twitter's role in increasing the visibility and expressions of resistance to the normative sexual order compensates for the lack of, or fear of, offline visibility in the Nigerian context (Onanuga, 2022).

Homophobia and misogynistic violence

In the dataset, expressions of violent misogyny, explicit lesbophobia and homophobia were implicit in the hateful outrage expressed around queer feminists and their allies. However, their vulnerable position was a potential source of strength from which they negotiated different ways of being visible and transgressive. X users claiming to want to learn more about lesbian sexual practices wrote cynical comments to create an atmosphere of amusement at the denigration of queer users. Jane (2014) has argued that inequalities on social media are not reducible to access but to the ability to feel safe or able to speak without receiving abusive or objectifying comments. There were accusations of lesbian practices of Nigerian women destroying the Nigerian Igbo culture. These tweets also claimed religious victimhood and expressed their disappointment with allies for supporting offensive lesbian practices that were an abomination to African tradition and Christian practices. This sentiment was fuelled by the belief that homosexuality was undermining African traditions, while others outrightly used culture and religion to stoke misogynistic sentiments and transphobic slurs when addressing sexual minorities on X. Interestingly, not all of the homophobic tweets towards lesbian women came from heterosexual men. Many such tweets came from a self-proclaimed feminist cis woman expressing disgust about the lesbians' imagined sexual behaviour and expression. They view sexual relationships between lesbian women as inappropriate, strange or even gross.

Sample tweets posted in protest claimed African culture victimhood.

It's the thought of two naked ladies for me! Like hell no! The worst is them using a toy dick rather than getting the real thing from a human.

How do you do it in the bedroom? The man positions sometimes and then presses reverse.? Oya makes you explain. @Jude

The following tweet was retweeted several times:

No one is interested in your private life, you sick, mentally deranged lesbian women! Your obnoxious mindset is vile. What you practice is not feminism. It is nympho narcissism. Stop shoving it down our throats. @Stellamia

The visceral nature of these reactions indicates the ways queerness is repudiated and viewed as shameful and inappropriate. Not only heterosexual men hold negative attitudes towards lesbian women. Some women also joined in the expression of negative attitudes. It shows that the homophobic attitudes that some feminists hold are what, in part, keep heterosexual feminists and queer women from completely unifying.

This tendency was most pronounced among the men who wrote comments on posts by @UjuAnya and other members of the LGBTQ community. Some of the strategies used to confront homophobic and misogynistic attacks were the use of expletives, blocking, sarcasm and counter-attacks.

As Sobieraj (2020) has argued, abuse targeting women is often more than interpersonal bullying but is a visceral response to the threat of equality in digital conversations and arenas that men prefer to control.

Uju, go get yourself a man! Oshe! Why do you need another woman dressed up as a man. This is crazy

This woman has brought shame to Igbo people! You engage in despicable acts. Please stop associating with the Igbo culture. Igbo culture forbids lesbianism.

Shameless lesbian! May your children grow up to hate you for the bad example you set for them @Abim

Figure 2.3 Homophobic comments on @UjuAnya's tweets

That woman is no hero is rude, highly offensive, and without manners. May God deliver you.

What these tweets have in common, aside from the misogynistic and homophobic slurs, is that they perceive sexual minorities as un-Nigerian/un-African, un-Godly and a threat to Nigerian culture and society. This idea that homosexuality is un-African has been a subject of intense debate in African feminist discourses. It is a trope, a transphobic and misogynistic attack against Anya because of her sexual identity, race and Nigerian identity. In feminist theorizing, scholars show how intersecting oppressions create institutions where injustice is imposed even further on those with intersecting identities. For example, Lorde (2012) suggests that Black women's oppression comes as a result of their blackness and 'womanness'. It corroborates Okech's (2021) argument about how tropes such as the angry, unhappy Black are used to order and silence women. In refusing to be silenced in this instance, Anya and her allies managed to flip the narratives in many cases to advocate for recognition of LGBTQ people and their social inclusion and to push back against discrimination practices.

The discourses around #UjuAnya highlighted instances of resistance and resilience manifesting on X. Divided responses to the demand for justice and feminists' criticism of government elites and clergies demonstrated the relevance of acknowledging how monolithic conceptions of culture and religion reinforce women's oppression, and Nigerian feminists' insistence of being heard. It also shows how misogyny and homophobia are effective tools for the repression and subjugation of feminists online. Although the hashtags never resulted in street marches, unapologetic self-assertion and stoic resilience are positive developments in queer women's defiance against prevailing social norms, restrictions and disempowerment.

Despite the hostile Nigerian X environment, queer identities constantly demonstrate the willingness to be affronted and to engage in the struggles of feminism with resilience, even in the face of ad hominem attacks. Queerness is a contemporary cultural struggle in a fledgling democracy that intersects with multiple struggles of exclusion in more conventional forms of citizenship. Reconceptualizing feminist citizenship as non-essentialist, and non-binary, and positioning it as a composite of various subjectivities, affiliations and hierarchies forms the basis for a radical approach that cultivates resilience geared towards progressive feminist citizenship and gender relations (Isin and Ruppert, 2020; Emejulu and McGregor, 2019).

In alignment with Isin and Ruppert's (2020) ideas of becoming digital subjects of power rather than being subjected to control, the strategies queer feminists employed in responding to many homophobic and misogynistic attacks on X include ignoring, educating, blocking and also the use of expletives to fight back. Through their activism and resistance, feminists on Nigeria's X challenged the effectiveness of online misogyny as a tool used to subjugate and silence women, rendering it ineffective. This refusal to cower in the face of online threat, abuse and hostilities means that sexual minorities are no longer powerless victims. They are reclaiming the social space by articulating their political subjectivity, and this process may in turn reshape public discourse and attitudes.

Figure 2.4 Homophobic tweets on X

Conclusion

This chapter has explored the tensions surrounding queer women's sexual identity on Nigeria's X, especially the conflation of the lack of propriety and ungodliness ascribed to lesbian women and the contention with what constitutes good citizenship. It demonstrated how the hashtag #UjuAnya provided a platform where @UjuAnya and other non-binary individuals and their allies claimed their rights to queer feminist digital citizenship and how the process of their political subjectivation involved tactics of resilience and defiance, which worked as agentic strategies in challenging existing scripts that have continued to reproduce their exclusion and disengagement from full citizenship.

Given that detractors and anti-feminists deployed violent homophobic onslaughts and digital misogynistic violence to curtail their resistance, queer feminists entering Nigeria's X were doubly defiant. In the comments, @UjuAnya's ability and boldness to be openly lesbian was often attributed to the freedom she enjoys living in the diaspora. But her use of X to advocate for LGBTQ rights was also transgressive, even within conservative Western social contexts. Being openly lesbian and advocating publicly for LGBTQ rights is not something every woman dares to do. It is a public performance of gender and sexuality that challenges notions of hegemonic masculinity in spaces men would prefer to control. In fighting back, queer feminists exploited the counter-narrative affordances of X, such as the block button, retweet and irreverent speech acts. Their use of expletives to counter misogynistic abuse was also regarded as a transgression of the boundaries of propriety for Nigerians.

Being forced to be at the intersection of various structural oppressions, the bottom-up resistance, resilience and visibility #UjuAnya inspired on the platform opened the space for others to vocalize their feminist and sexual identities and their oppression without fear of intimidation. Ultimately, trans and lesbian women on Nigeria's X can be true to their identity and sexuality and be subversive rather than being submissive to forces that control and undermine their identity. Despite the deluge of digital misogyny and homophobia, X plays an important role in helping Nigeria's LGBTQ community reveal their authentic selves when a community of defiant feminists work together, allowing for the creation of new forms of identity and agency. However, responses and interactions with heterosexual individuals tended to be more kneejerk and confrontational as opposed to discursive interactions that could change hearts and minds.

Notes

1. 'Queer' is used here as an umbrella term to denote a diversity of identities that challenge the binary categories of female and male. Hence, queer is a marker of non-normative identity, and queering as different modes of doing gender and sexuality politics.
2. At time of writing there was significant outcry and contention surrounding the legitimacy of same-sex relationships, owing to Donald Trump's decree that the US

federal government will acknowledge only two sexes. It shows how political homophobia is reinforced through hegemonical, heteropatriarchal discourses that validate the othering of queer sexualities.

3   When data was collected for this study in 2022, the platform X was still called Twitter. For this reason, I use X and Twitter interchangeably in this chapter. Twitonomy is a web analytic tool specifically for collecting data and tracking statistics on the platform. It is still in use, though it only offers a premium plan. When the first draft was written, it had a free version.

Bibliography

Adomi, E. E. (2005), 'Internet Development and Connectivity in Nigeria', *Program*, 39 (3): 257–68.

Akanji, O., Epprecht, M. and Nyeck, S. (2013), 'Human Rights Challenge in Africa: Sexual Minority Rights and the African Charter on Human and Peoples' Rights', in S. N. Nyeck and M. Epprecht, *Sexual Diversity in Africa: Politics, Theory, Citizenship*, Montreal: McGill-Queen's University Press,19–36.

Akinbobola, Y. (2019), 'Neoliberal Feminism in Africa', *Soundings*, 71 (Spring): 50–61.

Akinbobola, Y. (2020), 'Defining African Feminism(s) While #BeingFemaleinNigeria', *African Diaspora*, 12 (1-2): 64–88.

Akpojivi, U. (2019), 'I Won't be Silent Anymore: Hashtag Activism in Nigeria', *Communicatio: South African Journal of Communication Theory and Research*, 45 (4): 19–43.

Alichie, B. (2023), 'Communication at the Margins: Online Homophobia from the Perspectives of LGBTQ+ Social Media Users', *Journal of Human Rights*, 22 (3): 269–83.

Alkali, A. U., Jimeta, U., Magashi, A. I. and Buba, T. M. (2014), 'Nature and Sources of Nigerian Legal System: An Exorcism of Wrong Notion', *International Journal of Business, Economics and Law*, 5 (4): 1–10.

Armisen, M. (2016). *We Exist: Mapping LGBTQ Organizing in West Africa*, Astraea Lesbian Foundation for Justice.

Ayodele Onanuga, P. (2020), 'Queer Nigerian Twitter Can Challenge Homophobia and Assert Sexual Agency', *Africa at LSE*. Retrieved from https://blogs.lse.ac.uk/africaatlse/2020/10/29/queer-nigerian-twitter-social-media-challenge-homophobia-and-assert-sexual-agency/.

Baer, H. (2018). 'Redoing Feminism: Digital Activism, Body Politics, and Neoliberalism', *Digital Feminisms*, Routledge.

Bailey, M. (2021), 'Misogynoir Transformed: Black Women's Digital Resistance', *Misogynoir Transformed*, New York: New York University Press.

Baumgardner, J. (2011) 'Is There a Fourth Wave? Does it Matter?', *Feminist.com*. Retrieved from https://www.feminist.com/resources/artspeech/genwom/baumgardner2011.html

Baxter, P. (2015), *Biafra: The Nigerian Civil War 1967–1970*, Warwick, UK: Helion and Company.

Berents, H. (2016), 'Hashtagging Girlhood: #IAmMalala, #BringBackOurGirls and Gendering Representations of Global Politics', *International Feminist Journal of Politics*, 18 (4): 513–27.

Biddie, S. (2022), 'Twitter Censored Professor's Post for "Abusive Behaviour" Toward the Queen', *The Intercept*, 9 September. Available at https://theintercept.com/2022/09/09/queen-dead-twitter-censor-abuse-uju-anya/.

Bosia, M. J. and Weiss, M. L. (2013), 'Political Homophobia in Comparative Perspective', *Global Homophobia: States, Movements, and the Politics of Oppression*, Champaign, IL: University of Illinois Press, 1–29.

Braun, V. and Clarke, V. (2021), 'One Size Fits All? What Counts as Quality Practice in (Reflexive) Thematic Analysis?', *Qualitative Research in Psychology*, 18 (1): 328–52.

Brinkel, T. and Ait-Hida, S. (2012), 'Boko Haram and Jihad in Nigeria', *Scientia Militaria: South African Journal of Military Studies*, 40 (2): 1–21.

Buckle, L. (2020), 'African Sexuality and the Legacy of Imported Homophobia', *Stonewall*, 1 October 2020. Available from: https://www.stonewall.org.uk/news/african-sexuality-and-legacy-imported-homophobia.

Caruth, C. (2016), *Unclaimed experience: Trauma, narrative, and history*, Baltimore, MD: JHU Press.

Chamberlain, P. (2016), 'Affective Temporality: Towards a Fourth Wave', *Gender and Education*, 28 (3): 458–64.

Chemaly, S. (2018), *Rage Becomes Her*, New York: Simon and Schuster.

Clark-Parsons, R. (2022), *Networked Feminism: How Digital Media Makers Transformed Gender Justice Movements*, Oakland, CA: University of California Press.

Clode, L. (2018), 'A Defence of Fourth Wave Feminism', *Gender, Place and Culture*, 22 August 2018. Retrieved from https://genderplaceandculture.wordpress.com/2018/08/22/post-9-of-gpc25-a-defence-of-fourth-wave-feminism-by-lucy-clode/.

Cochrane, K. (2013), *All the Rebel Women: The Rise of the Fourth Wave of Feminism*, London: Guardian Books.

Collins, P. H. (2015), 'Intersectionality's Definitional Dilemmas', *Annual Review of Sociology*, 41 (1): 1–20.

Collins, P. H. (2022), *Black Feminist Thought: Knowledge, Consciousness, and the Politics of Empowerment*, New York and London: Routledge.

Couldry, N. (2010), *Why Voice Matters: Culture and Politics after Neoliberalism*, London: Sage Publications.

Crenshaw, K. (1989), 'Demarginalizing the Intersection of Race and Sex: A Black Feminist Critique of Antidiscrimination Doctrine, Feminist Theory and Antiracist Politics', *The University of Chicago Legal Forum*, 140: 139–67.

Crenshaw, K. (1991), 'Mapping the Margins: Intersectionality, Identity Politics, and Violence Against Women of Color', *Stanford Law Review*, 43 (6): 1241–99.

Dambo, T. H., Ersoy, M., Auwal, A. M., Olorunsola, V. O. and Saydam, M. B. (2022), 'Office of the Citizen: A Qualitative Analysis of Twitter Activity During the Lekki Shooting in Nigeria's #EndSARS Protests', *Information, Communication & Society*, 25 (15): 2246–63.

Davies, E. B. (2018), 'Third Wave Feminism and Transgender', *Strength Through Diversity*, New York and London: Routledge.

Diamond, L. (1988), *Class, Ethnicity, and Democracy in Nigeria: The Failure of the First Republic*. Syracuse: Syracuse University Press.

Dosekun, S. (2022), 'The Problems and Intersectional Politics of "#BeingFemaleinNigeria"', *Feminist Media Studies*, 4: 1–17.

Drüeke, R. and Zobl, E. (2018), 'Online Feminist Protest Against Sexism: The German-Language Hashtag #aufschrei', in C. Scharff, C. Smith-Prei and M. Stehle (eds) *Digital Feminisms*, London and New York: Routledge.

Duffy, B. E. and Meisner, C. (2023), 'Platform Governance at the Margins: Social Media Creators' Experiences with Algorithmic (In)visibility', *Media, Culture & Society*, 45 (4): 285–304.

Emejulu, A. and McGregor, C. (2019), 'Towards a Radical Digital Citizenship in Digital Education', *Critical Studies in Education*, 60 (2): 131–47.

Epprecht, M. and Egya, S. E. (2011), 'Teaching about Homosexualities to Nigerian University Students: A Report from the Field', *Gender and Education*, 23 (4): 367–83.

Falola, T. and Heaton, M. M. (2008), *A History of Nigeria*, Westport, CT: Cambridge University Press.

Faniyi, O. (2023), 'Intersectionality in/through Nigeria's Feminist Hashtag Activism', *Communication, Culture & Critique*, 16 (4): 110–12.

Faniyi, O., Nduka-Nwosu, A. and Gajjala, R. (2023), '#SayHerNameNigeria: Nigerian Feminists Resist Police Sexual Violence on Women's Bodies', in B. Wiens, M. MacArthur, S. MacDonald and M. Radzikowska (eds), *Stories of Feminist Protest and Resistance: Digital Performative Assemblies*, Lanham, MD: Lexington Books.

Ferree, M. M. and Roth, S. (1998), Gender, Class, and the Interaction Between Social Movements: A Strike of West Berlin Day Care Workers', *Gender & Society*, 12 (6): 626–48.

Fotopoulou, A. (2012), 'Intersectionality Queer Studies and Hybridity: Methodological Frameworks for Social Research', *Journal of International Women's Studies*, 13 (2): 19–32.

Fotopoulou, A. (2017), *Feminist Activism and Digital Networks: Between Empowerment and Vulnerability*, London: Palgrave MacMillan.

Foucault, M. (1978), *The History of Sexuality: Volume I: An Introduction*, New York: Pantheon Books.

Gill, R. and Orgad, S. (2018), 'The Shifting Terrain of Sex and Power: From the "Sexualization of Culture" to #MeToo', *Sexualities*, 21 (2): 1313–24.

Hanckel, B., Vivienne, S., Byron, P., Robards, B. and Churchill, B. (2019), '"That's Not Necessarily for Them": LGBTIQ+ Young People, Social Media Platform Affordances and Identity Curation', *Media, Culture & Society*, 41 (1): 1261–78.

Hayward, F. M. ed. (2019), *Elections in Independent Africa*, London and New York: Routledge.

Hazzad, A. (2022), 'Nigerian Islamic Court Orders Death by Stoning for Men Convicted of Homosexuality', *Reuters*, 2 July. Available at https://www.reuters.com/world/africa/nigerian-islamic-court-orders-death-by-stoning-men-convicted-homosexuality-2022-07-02/.

Hildebrandt, T. and Chua, L. J. (2017), 'Negotiating In/visibility: The Political Economy of Lesbian Activism and Rights Advocacy', *Development and Change*, 48 (3): 639–62.

Hines, S. (2005), '"I am a Feminist but … ": Transgender Men and Women and Feminism', in J. Reger (ed.), *Different Wavelengths*, New York and London: Routledge.

hooks, b. (1981), *Ain't I A Woman: Black Women and Feminism*, Boston: South End Press.

Isin, E. and Ruppert, E. (2015), *Being Digital Citizens*, London: Rowman & Littlefield.

Isin, E. and Ruppert, E. (2020), *Being Digital Citizens*, London: Rowman & Littlefield.

Jackson, S. J., Bailey, M. and Foucault Welles, B. (2018), '#GirlsLikeUs: Trans Advocacy and Community Building Online', *New Media & Society*, 20 (5): 1868–88.

Jackson, S. J., Bailey, M. and Welles, B. F. (2020), *#HashtagActivism: Networks of Race and Gender Justice*, Cambridge, MA: MIT Press.

Jane, E. A. (2014), '"Your a Ugly, Whorish, Slut". Understanding E-bile', *Feminist Media Studies*, 14 (4): 531–46.

Job, P. (2020), '#EndSARS is a Huge Moment in Nigeria's Queer History', *Washington Post*, 5 November 2020.

Johns, A. and McCosker, A. (2015), 'Social Media Conflict: Platforms for Racial Vilification, or Acts of Provocation and Citizenship?', *Communication, Politics & Culture*, 47 (3): 44–54.

Kemp, S. (2024), 'Digital 2024: Nigeria', *DataReportal*, 23 February. Available at https://datareportal.com/reports/digital-2024-nigeria#:~:text=Internet%20use%20in%20Nigeria%20in,on%20data%20for%20further%20details.

Lardy, M. (2022), 'Bisi Alimi Shares Throwback Video of Labour Party VP Candidate, Yusuf Datti Baba-Ahmed, Calling for the Killing of Homosexuals', *Linda Ikeje's Blog*, 9 September.

Leurs, K. and Ponzanesi, S. (2011), 'Mediated Crossroads: Youthful Digital Diasporas', *M/C Journal*, 14 (2).

Linabary, J. R., Corple, D. J. and Cooky, C. (2019), 'Feminist Activism in Digital Space: Postfeminist Contradictions in #WhyIStayed', *New Media & Society*, 22 (10): 1827–48.

Lorde, A. (2012), *Sister Outsider: Essays and Speeches*, Berkeley, CA: Crossing Press.

Loza, S. (2014), 'Hashtag Feminism, #SolidarityIsForWhiteWomen, and the Other #FemFuture', *Ada: A Journal of Gender, New Media, and Technology*, 5.

Madunagu, B. E. (2008), 'The Nigerian Feminist Movement: Lessons from Women in Nigeria, WIN', *Review of African Political Economy*, 35 (118): 666–73.

Maragh-Lloyd, R. (2020), 'Civic Debate and Self-care: Black Women's Community Care Online', in G. Bouvier and J. Rosenbaum (eds), *Twitter, the Public Sphere, and the Chaos of Online Deliberation*, Cham, CH: Springer.

McCosker, A., Vivienne, S. and Johns, A. (2016), *Negotiating Digital Citizenship: Control, Contest and Culture*, London: Rowman & Littlefield.

Megarry, J. (2018), 'Under the Watchful Eyes of Men: Theorising the Implications of Male Surveillance Practices for Feminist Activism on Social Media', *Feminist Media Studies*, 18 (2): 1070–85.

Megarry, J. (2020a), 'A Fourth Wave or a Fool's Errand?', *The Limitations of Social Media Feminism: No Space of Our Own*, Cham: Springer, 1–34.

Megarry, J. (2020b), '"On the Internet, There is No Women-Only Space": Male Power in Digital Networks', *The Limitations of Social Media Feminism: No Space of Our Own*, Cham: Springer.

Mohanty, C. (1988), 'Under Western Eyes: Feminist Scholarship and Colonial Discourses', *Feminist Review*, 30 (1): 61–88.

Mossberger, K., Tolbert, C. J. and McNeal, R. S. (2007), *Digital Citizenship: The Internet, Society, and Participation*, Cambridge, MA: MIT Press.

Munro, E. (2013), 'Feminism: A Fourth Wave?', *Political insight*, 4 (2): 22–5.

Murthy, D. (2013), *Social Communication in the Twitter Age*, Cambridge, UK: Polity Press Cambridge.

Mwangi, E. (2014), 'Queer Agency in Kenya's Digital Media', *African Studies Review*, 57 (2), 93–113.

Nanditha, N. (2022), 'Exclusion in #MeToo India: Rethinking Inclusivity and Intersectionality in Indian Digital Feminist Movements', *Feminist Media Studies*, 22 (14): 1673–94.

Nartey, M. (2022), 'Centering Marginalized Voices: A Discourse Analytic Study of the Black Lives Matter Movement on Twitter', in *Voice, Agency and Resistance*, London: Routledge.

Njoroge, D. (2016), 'Broken Silence: #Bringbackourgirls and the Feminism Discourse in Nigeria', in B. Mutsvairo (ed.) *Digital Activism in the Social Media Era: Critical*

Reflections on Emerging Trends in Sub-Saharan Africa, Cham: Palgrave MacMillan, 311–26.

Nkordeh, N., Bob-Manuel, I. and Olowononi, F. (2017), 'The Nigerian Telecommunication Industry: Analysis of the First Fifteen Years of the Growths and Challenges in the GSM Market (2001–2016)', *World Congress on Engineering and Computer Science*, San Francisco, US, 25–7.

Novak, A. (2016), *Defining Identity and the Changing Scope of Culture in the Digital Age*. Hershey, PA: IGI Global.

Nwabunnia, O. A. (2021), '#EndSARS Movement in Nigeria: Tensions and Solidarities Amongst Protesters', *Gender & Development*, 29 (2-3): 351–67.

Nyabola, N. (2018), 'Kenyan Feminisms in the Digital Age', *Women's Studies Quarterly*, 46 (3&4): 261–72.

Nyanzi, S. (2014), 'Queer Pride and Protest: A Reading of the Bodies at Uganda's First Gay Beach Pride', *Signs: Journal of Women in Culture and Society*, 40 (1): 36–40.

O'Flaherty, M. and Fisher, J. (2008), 'Sexual Orientation, Gender Identity and International Human Rights Law: Contextualising the Yogyakarta Principles', *Human Rights Law Review*, 8 (2): 207–48.

Obadare, E. (2015), 'Sex, Citizenship and the State in Nigeria: Islam, Christianity and Emergent Struggles over Intimacy', *Review of African Political Economy*, 42 (143): 62–76.

Obadare, E. (2021), 'A Hashtag Revolution in Nigeria', *Current History*, 120 (826): 183–8.

Obia, V. (2020), '#EndSARS, a Unique Twittersphere and Social Media Regulation in Nigeria', *LSE Blogs*, 11 November 2020. Retrieved from https://blogs.lse.ac.uk/medialse/2020/11/11/endsars-a-unique-twittersphere-and-social-media-regulation-in-nigeria/.

Odili, N. (2021), 'Awareness of the Use and Impact of Facebook and WhatsApp Among Undergraduate Students in Tertiary Institutions: A Review of the Literature', *MiddleBelt Journal of Library and Information Science*, 19.

Ogundipe-Leslie, M. (1994), *Re-creating Ourselves: African Women & Critical Transformations*, Trenton, NJ: Africa World Press.

Okech, A. (2021), 'Feminist Digital Counterpublics: Challenging Femicide in Kenya and South Africa', *Signs: Journal of Women in Culture and Society*, 46 (4): 1013–33.

Oladapo, O. A. and Ojebuyi, B. R. (2017), 'Nature and Outcome of Nigeria's #NoToSocialMediaBill Twitter Protest against the Frivolous Petitions Bill 2015', in O. Nelson, B. Ojebuyi and A. Salawu (eds), *Impacts of the Media on African Socio-Economic Development*, Hershey, PA: IGI Global.

Olofintuade, A. (2017), Female in Nigeria: Profile, *Feminist Africa*, 22, 163–73.

Onanuga, P. A. (2021), '#ArewaAgainstLGBTQ Discourse: A Vent for Anti-homonationalist Ideology in Nigerian Twittersphere?', *African Identities*, 21 (4): 1–23.

Onanuga, P. A. (2022), 'Navigating Homophobia and Reinventing the Self: An Analysis of Nigerian Digital Pro-gay Discourse'. *Gender & Language*, 16 (1).

Osibu, P. (2022), *The Challenges of Navigating Feminism in Nigeria: A Case for Third Space Feminism*, Champaign, IL: Illinois State University.

Osinubi, T. A. (2016), 'Queer Prolepsis and the Sexual Commons: An Introduction', *Research in African Literatures*, 47 (2): vii–xxiii.

Owoh, U. O. (2020), 'Queer Nigerians Face Police Brutality. Why Were They Erased From #EndSARS?', *Open Democracy*, 30 December 2020, https://www.opendemocracy.net/en/5050/queer-nigerians-face-police-brutality-why-were-they-erased-endsars/.

Patrick, C. E. F. (2016), 'LGBT Rights Movement in Africa and the Myth of the Whiteman's Superiority', *Journal of Globalization Studies*, 7, 139–51.

Pereira, C. and Ibrahim, J. (2010), 'On the Bodies of Women: The Common Ground Between Islam and Christianity in Nigeria', *Third World Quarterly*, 31 (6): 921–37.

Pindi, G. N. (2020), 'Beyond Labels: Envisioning an Alliance Between African Feminism and Queer Theory for the Empowerment of African Sexual Minorities Within and Beyond Africa', *Women's Studies in Communication*, 43 (1): 106–12.

Rogers, R. (2019), *Doing Digital Methods*, London: Sage.

Salami, M. (2018), 'Feminism in Nigeria – By and For Who? To What Extent Does Contemporary Nigerian Feminism Reflect Nigerian Women's realities?', *LuXemburg*, September 2018, Available from: https://zeitschrift-luxemburg.de/artikel/feminism-in-nigeria-by-and-for-who/.

Sobieraj, S. (2020), *Credible Threat: Attacks Against Women Online and the Future of Democracy*, New York: Oxford University Press.

Sogunro, A. (2018), 'Citizenship in the Shadows', *College Literature*, 45 (4): 632–40.

Sogunro, A. (2022), 'An Analysis of Political Homophobia, Elitism and Social Exclusion in the Colonial Origins of Anti-gay Laws in Nigeria', *African Human Rights Law Journal*, 22, 493–519.

Strand, C. (2019), 'Navigating Precarious Visibility: Ugandan Sexual Minorities on Twitter', *Journal of African Media Studies*, 11 (2): 229–56.

Sunday, O. (2020), 'In One of the Most Religious Countries in the World, Nigerian Churches and Mosques have Joined the Fight Against the Coronavirus', *Equal Times*, 27 May 2020. Retrieved from https://www.equaltimes.org/in-one-of-the-most-religious?lang=en.

Tamale, S. (2011), 'Researching and Theorising Sexualities in Africa', in *African Sexualities: A Reader*, Pambazuka Press, 11–36.

Tamale, S. (2020), *Decolonization and Afro Feminism*, Ottawa: Daraja Press.

Toyo, N. and Adegbeye, O. (2021), 'Are Different Generations of Nigerian Feminists Ready to Join Forces? An Interview', *Heinrich-Böll-Stiftung*, 8 July 2021.

Uwalaka, T. (2022), 'Social Media as Solidarity Vehicle During the 2020 #EndSARS Protests in Nigeria', *Journal of Asian and African Studies*, 59 (2): 00219096221108737.

Uwazuruike, A. R. (2020), '#EndSARS: The Movement Against Police Brutality in Nigeria', *Harvard Human Rights Journal*, 33.

Vickery, A. E. (2018), 'After the March, What? Rethinking How We Teach the Feminist Movement', *Social Studies Research and Practice*, 13 (3): 402–11.

Watch, H. R. (2013), 'This Alien Legacy The Origins of "Sodomy" Laws in British Colonialism', *Human Rights Watch*, 17 December 2008. Available from https://www.hrw.org/report/2008/12/17/alien-legacy/origins-sodomy-laws-british-colonialism.

Yeku, J. (2020), '#endsars Shows Why the Police is Not 'Your Friend', *Africa at LSE*, 12 November 2020. Available at https://blogs.lse.ac.uk/africaatlse/2020/11/12/endsars-shows-why-the-police-is-not-your-friend/.

Yue, A. and Lim, R. P. (2022), 'Digital Sexual Citizenship and LGBT Young People's Platform Use', *International Communication Gazette*, 84 (4): 331–48.

Yuval-Davis, N. (2006), 'Intersectionality and Feminist Politics', *European Journal of Women's Studies*, 13 (3): 193–209.

Zimmerman, T. (2017), '#Intersectionality: The Fourth Wave Feminist Twitter Community', *Atlantis: Critical Studies in Gender, Culture & Social Justice*, 38 (1): 54–70.

Zivi, K. (2012), *Making Rights Claims: A Practice of Democratic Citizenship*, New York: Oxford University Press.

Chapter 3

'OF COURSE WE ARE ANGRY': LUSAKA WOMEN AND THE ZAMBIAN FEMINISTS FACEBOOK PAGE

Chishimba Kasanga and Priscilla Boshoff

The internet has enabled a global feminist community to engage in discussion and activism, with campaigns like #MeToo and #AmINext garnering widespread online support from women worldwide. Despite this wave of attention, feminism remains controversial in some parts of Africa. This is the case in Zambia, where the Zambian Feminists Facebook page challenges patriarchy and gender norms in a highly heteronormative society. Founded in 2018 by a feminist activist in Lusaka, the page addresses gender-based violence, rape, LGBTQI issues and patriarchal practices, aiming to bring these issues to public attention and foster discussion among women to inspire change.

In this chapter we use digital ethnography, content analysis and interviews to give a snapshot of the feminist actions and citizenship inspired by the Zambian Feminists Facebook page. To this end, the Zambian Feminists Facebook page explicitly addresses a community of Zambian women who identify as feminists. Speaking on behalf of these followers, the stance of the page is evident in its tagline: 'of course we are angry'. It describes Zambian women's anger at not 'being heard' and attempts to provide a space where women can listen to each other and 'share a voice'. The page posts material concerning, *inter alia*, patriarchal practices, LGBTQI issues, gender-based violence, rape and menstrual hygiene, often using examples from local media to prompt discussion. The provocative content and ensuing debates quickly boosted the page's popularity, amassing about 32,000 followers at the time that this research was conducted in 2020.

This chapter contributes to the current debates around the growing popularity and effectiveness of online feminism and the consequences of this for a sense of citizenship. Central to our enquiry is the question posed by Roberts and Bosch (2023: 20): 'under what conditions [is it] possible to harness the positive affordances of digital technology for citizenship and social change?' The 'conditions' in this case are established by the specific context of Lusaka, Zambia's capital city. Lusaka, like many African cities, is characterized by contesting discourses regarding appropriate gender roles for our contemporary times. Here, customary, Christian and Western ideas and practices jostle each other, often uncomfortably, as people negotiate the boundaries of acceptable male and female behaviour and the social and cultural values associated with them. These varying discourses and the specific urban context in which they circulate provide the stage on which the followers of the Zambian Feminists Facebook page construct their feminist identities and seek new, more just and equitable forms of social and cultural belonging – of citizenship.

With this stage in mind, we ask: 'What meanings do women make of the Zambian Feminists content and comments, and in what ways are they able to put these meanings into practice within their everyday lives as self-proclaimed feminists?' We follow Nyamnjoh's (2006) view that citizenship is an ongoing process rather than a final goal. Similarly, gender relations, as part of citizenship, continuously evolve as we redefine social norms related to the identities associated with our reproductive capacity (Connell, 2009). Gender serves as a lens, mediating social experiences and shaping our place in the social order. While gender norms can guide behaviour, they can also limit alternative understandings and ways of being. Feminism challenges these restrictions, promoting flexibility and enhancing democratic citizenship by opposing exclusionary practices and creating new spaces for belonging and association.

Taking this understanding of feminism as a departure point, our aim in this chapter is not to simply point out the deficits and inequalities of gender relations in Lusaka. Nor is it to recount the hurdles that cisgender women face as they attempt to create a more inclusive and equitable society. Rather, the aim is to appreciate the context in which these dynamics play out and how they shape the possibilities for feminist action and citizenship. The Zambian Feminists Facebook page provides a small window into this fraught arena, allowing us to glimpse what issues are at stake, what kinds of actions and actors are deplored, and what alternative values and behaviours are desired. It also helps us to appreciate the strategies adopted by women who identify as feminists online but struggle to put their feminist values into action at home or in the workplace.

We first set the context of the research, starting with a description of the evolution of Zambia's gender relations. With this foundation established, we move on to discuss the scholarship that helps us to frame and understand the Zambian Feminists Facebook content. Here, we draw on Connell (2009), whose conceptualization of gender relations helps to account for the dynamics of gender in Zambia's post-colonial era. We then explain our choice of research methods before presenting the findings from our textual and interview data, and we end by arguing that feminism, as seen on the Zambian Feminists Facebook page, confronts local patriarchal constraints by advocating for more fluid gender definitions and strengthening democratic citizenship through resistance to exclusionary practices while creating inclusive spaces for belonging and connection.

Gender relations in Zambia: A complex history

Ray (2018) reminds us that all gender relations are the result of history. This history must be taken into account if we are to truly understand and appreciate the gender relations that characterize our contemporary milieu. The gender relations that we encounter in Africa today have emerged from the complex experiences of the impact of colonization on indigenous social systems and the postcolonial political and social dispensations that replaced colonial rule. How Zambian women 'belong' within contemporary Zambian society needs to be understood in relation to this turbulent history, in which indigenous gender relations were deliberately dismantled (Connell, 2009) and remade to suit the demands of colonial and postcolonial society. We can periodize this history, starting in precolonial times, through the period of colonial control and culminating in the country's post-independence insertion into the global economic and cultural order.

Precolonial gender relations in Zambia depended on whether one belonged to a matrilineal or patrilineal clan. Most ethnic groups were matrilineal, where women held significant social and economic status (Parpart, 1991; Crehan, 1997; Taylor, 2006). A girl's birth was celebrated, as her husband's labour benefited her clan, and her children increased her family's numbers, granting women considerable autonomy regarding residence and childbearing (Richards, 1940). Divorce was straightforward due to the absence of bridewealth. In patrilineal clans, women had less autonomy, as bridewealth transferred control of their reproductive and labour power to the husband's family (Parpart, 1991; 1994). Bilateral lineages focused on spousal rights, allowing flexibility in residence and property inheritance from both parents (Parpart, 1991; 1994).

Customarily, both boys and girls were socialized into distinct gender roles, which prepared them for adult responsibilities. These included the skills of hunting for men and agricultural production for women (Rasing, 2001; Allen, 2010). Boys and girls also underwent initiation rites that prepared them for adulthood, sex and marriage. A girl's initiation rites took place after her first menstruation, *Ichisungu* (Rasing, 2001). These rites marked the passage from childhood to womanhood and emphasized the girl's reproductive roles within marriage, her domestic and agricultural duties, and respect for her elders and her future husband, as well as addressing sexuality and food taboos (Rasing, 2001). Having undergone initiation, a girl would be ready for marriage; polygyny was common (Taylor, 2006). This brief sketch suggests that people in precolonial Zambia had diverse gender relations. Women in matrilineal lineages held significant status and played important roles, though they were ultimately subordinate to men. These customs faced intense pressure after Zambia's colonization in 1888, as it transitioned from an agricultural society to a capitalist economy based on copper extraction. The colonial authorities and Christian missionaries restructured gender roles to align with modern capitalist life, promoting wage labour, urbanization and Christian monogamy (Evans, 2015).

Within this new milieu, men were taxed in order to compel them to work in mines or cities, while women were expected to stay in rural areas. However, in the course of the twentieth century, many women moved to urban spaces, becoming independent through trading, gardening, brewing beer and renting accommodation (Parpart, 1986). This shift increased women's desire for independence, but unmarried, economically independent women were seen as a threat to urban order (Ault, 1983: 183). By the 1940s and 50s, their access to cities was heavily restricted by social and legal sanctions imposed by colonial administrators, mine owners and customary patriarchs (Evans, 2015; Siwila, 2017).

In this way, under the tutelage of the mission schools and with the connivance of both customary leaders and colonial authorities, most women lost their relative autonomy and social status. They were directed towards new Western Christian norms of submissive, dependent, chaste and monogamous marital relations, domesticity and childbearing, attributes that became increasingly essential for social status and respectability in a rapidly modernizing Zambia (Epstein, 1981; Ferguson, 1999; Siwila, 2017). Following independence, the crash of the copper economy in the 1970s and neoliberal economic 'restructuring' in the 1990s led to widespread unemployment for men (Ferguson, 1999). They lost their status as providers for their households, a cornerstone of masculinity within local patriarchal gender relations. Ironically, given the efforts to keep them at home, women needed to take up employment and small-scale trading to supplement household income – but now without the social recognition that had been afforded to these activities in the past (Ferguson, 1999).

Today, Zambian cisgender women continue to bear the burden of this cumulative and intersecting history of custom, colonialism and Christianity. However, policy development, NGO activities and educational interventions have begun to modify the more inflexible aspects of gender arrangements (Siachitema, 2010), at least in urban areas (Evans, 2018). Evans (2014) also notes a growing flexibility in gender roles that she attributes to the sustained economic crisis. She argues that economic constraints have gradually accustomed both men and women to the necessity of women participating in the economy. While most engage in informal trading of some kind, a small though increasing number are educated for professional employment, taking up – and thereby normalizing – women's participation in 'masculine' occupations and responsibilities outside of the domestic sphere.

Media penetration in Zambia

We can add to this scenario the social impact of the media. Until the 1990s, the few available media outlets had been state owned and controlled. This changed somewhat when the liberalizing policies of structural adjustment in the last two decades of the twentieth century drove an increase in privately owned local media, including newspapers and radio and television stations (Willems, 2013). Since then, the expansion of digital networks and mobile phone uptake have increased exposure to and interactions with a diversity of online content for some. The digital divide means that digital access remains low overall: statistics from the World Bank (2021) showed that in 2021, around the time that this research took place, only about 21 per cent of Zambia's population had internet access, perhaps due to high data costs and limited infrastructure.

To provide a broader perspective, Zambia had 6.51 million internet users in 2023, with only 3.55 million using social media, comprising about 17 per cent of the population (DataReportal, 2023). Social media is mainly accessed via mobile phones, which are steadily increasing in number: early 2024 saw an estimated 16.40 million active mobile phones, representing about 78.7 per cent of the population (DataReportal, 2023). Facebook is the most popular platform, with 3.55 million users at the start of 2023 (approximately 17 per cent of the population), 55.7 per cent of whom were men (DataReportal, 2023). Facebook's popularity can be partly attributed to its Free Basics app, first launched in Zambia in 2014 (Willems, 2016). Free Basics is a mobile app created by Facebook that allows users to access certain data-light websites and services without photos and videos, which can be browsed for free. However, Free Basics has faced criticism for digital colonialism, including violating net neutrality principles, inadequately serving local language needs, collecting extensive user data and prioritizing Western corporate content (Nothias, 2020; Solon, 2017).

Digital ethnography

Initially driven by curiosity about topics on the Zambian Feminists page, this study evolved into a two-year digital ethnographic research project. Digital ethnography serves as a powerful tool for feminist researchers to investigate women's online experiences and the digital environments shaping them. Participant observation, a key aspect of digital ethnography, involves the researcher fully engaging in the participants' online lives and activities. This approach allows the researcher not just to observe but also to experience the participants' world (Becker and Geer, 1957; Gill and Johnson, 2002). The main advantage of participant observation is its ability to immerse the researcher deeply into the social and cultural context being studied. Anthropologist Evans-Pritchard (1981) describes this role as a 'doubly marginal' person, positioned between their own society and the society being researched.

This method was ideal for our study, as it allowed us to actively engage in discussions with the provocative posts on the page by commenting, liking and sharing. As followers of the page, we identified the active Lusaka women fans by their top fan badges. We purposefully selected these frequent contributors and then used snowball sampling (Deacon et al., 1999), where initial contacts suggested other potential participants. These women had formed an online community, knowing each other through interactions on the page despite never meeting offline. For our study, we selected nineteen women fans as participants.

Analysis of Zambian Feminists Facebook posts

Given the wide range of topics on the Zambian Feminists page, we began by conducting a thematic analysis to understand the content of the page, its discussion-enhancing topics, writing styles, humour and debates. Since the page does not follow a specific posting schedule, we coded each post to identify topics and construct themes and sub-themes (Bryman, 2012). Our sampling frame included 19 posts out of the 173 (approximately 11 per cent) shared between 3 January and 29 December 2019. We had to narrow down the number of posts and focus on those that were directly linked to our research. We purposefully selected posts with the highest engagement (likes, comments and shares) and those reflective of public discourse, even if they were not the most popular. These posts best illustrated the topic commonly posted on the page and the kinds of responses they engender. Using an inductive approach, we found that this small sample size allowed for a richly textured understanding of the messages conveyed by the Zambian Feminists Facebook page. For this study, a small sample was ideal for eliciting nuanced and complex analyses, especially in the context of an online community.

This period (3 January to 29 December 2019) was significant because, over this year, the Zambian Feminists page administrator actively organized programmes to raise awareness on feminism and gender issues affecting women in Zambia. For example, the Lusaka Women's march (19 January 2019) and the Yaka feminist festival (25-27 April 2019) were organized in collaboration with feminists from Malawi, Zimbabwe and South Africa to share their experiences. These events were held in Lusaka, and women fans, including this study's participants, were in attendance. This was also the period wherein #AmINext? reached its peak, in September 2019, after the rise in femicides in South Africa (Lyster, 2019). The #AmINext? movement brought attention to the number of femicide cases in Zambia, as women took to social media to share their experiences; in addition, a media frenzy regarding LGBTQI rights arose that polarized the country.[1] During this period, followers of the page actively debated and defended their views on the platform. We closely monitored these discussions by reading all the posts and their associated comments.

Meaning making

This study explored how Lusaka women fans of the Zambian Feminists Facebook page interpret, negotiate and integrate meanings from its representations into their lives as feminists. Recognizing participants as co-producers of meaning (Jensen, 1988; Schrøder et al., 2003), we conducted an audience reception study using focus groups to understand the meanings these women derive from the posts and comments on the page. Focus groups provide a means to examine how people form and negotiate views about media content through social interactions (Lunt and Livingstone, 1996). We aimed to understand how our diverse participants interpreted the page's content in a group setting. Narrowing down from nineteen posts, we selected eight posts covering the topics of women's dress, cultural practices, womanhood, sexuality, body types, Christianity, LGBTQI rights and masculinity to stimulate debate. These themes reflect typical gender struggles Zambian women face. We split participants into groups of three to five to ensure lively, interactive discussions that accommodated all perspectives.

Individual in-depth interviews

We then used individual in-depth interviews to clarify points from focus group discussions and to examine individual attitudes and values not observable during our digital ethnography. While interviews provide context-specific findings that cannot be generalized, they offer particular insights into individual views or opinions (Bryman, 2012). We conducted six interviews: one with the Zambian Feminists Facebook page administrator and five with women fans of the page. Participants were purposively selected for their interesting opinions and enthusiasm from the focus groups. Pseudonyms are used in presenting our findings, including referring to the page creator as the Zambian Feminists page Administrator.

Findings

In this section, we explore the complexities of Lusaka women fans' identities in relation to the Zambian Feminists page, moving beyond the notion that fans agree with all content. Rather, we examine the page as an online feminist community, its connection with its women followers, and the Administrator's role as content creator, moderator and curator. We address three main questions: Do women contributors feel supported in expressing their views given Zambia's patriarchal context? How has the page influenced participants' understanding of gender politics? How do they perceive and manage their role as feminists in Zambia's gender-unequal context?

'Of course we are angry'

In the interviews, we asked our participants about their reactions to the content of the Zambian Feminists page and how they engage with it based on their worldviews and feminist values. Their responses revealed that followers negotiate and resonate in different ways with the content. Preferences are shaped by religious beliefs, socio-cultural factors and personal relationships within the feminist community. Followers can be categorized into three groups: those who agree with most content, those who contest it, and those who follow but engage selectively due to disagreements they have with aspects of the content. What unifies these women fans is that they all follow the page and consume Zambian Feminists, albeit selectively. The women fans attribute their judicious consumption of Zambian Feminists to the page Administrator being too aggressive, or lacking objectivity, as well as their partiality to and intolerance of different opinions. The page posts important topics for them, but the way that some of the content is presented has led them to be discriminating in their consumption.

One participant, Beauty, emphasizes the need for diverse voices on the page to avoid it becoming a personal blog. The page Administrator acknowledges this concern, highlighting efforts to include diverse contributions while avoiding making the page solely about her views.

> **Beauty:** The page is personal, it's like a personal blog, but if that page admin aims to build a community of feminists there needs to be room for some conversation. I would want to see more authors and more collaborators to get a much richer diverse opinion.

For participants like Beauty, the Zambian Feminists Facebook page is the Administrator's personal blog. For her, the page needs more contributors to give diverse opinions and views about feminism. It is not enough for her to only read one person's view which claims to represent a community of feminists. We took Beauty's critique into consideration and brought it up during our interview with the Administrator. She acknowledged the danger of only expressing her views on the platform but emphasized that some of the content she shares are contributions from other women:

> Periodically, I invite contributions to the page, which I edit to ensure legal accuracy before sharing. We provide a platform for women to share their stories anonymously or otherwise. I'm cautious about the page becoming centred around my voice alone. I am just one voice in an ocean of many, I want more women to share their views because [I] am not writing a playbook on feminism. I'm scared of people saying what Zambian Feminists says is golden, it is not. No, it's not!

In her statement, the Administrator acknowledges the risk of the page becoming overly focused on her own perspectives, emphasizing that her viewpoint is just one among many. She rejects any notion of playing a definitive role or having the only answer regarding feminism in Zambia, explicitly stating she is not attempting to create a definitive guide on the subject. By inviting other contributors to the page, she aims to foster a platform that embraces a wide range of viewpoints, stories and experiences that resonate with diverse women followers. This openness also allows her to learn from others, recognizing her own limitations in understanding the diversity of feminist concerns and women's issues in Zambia. Yet, despite these efforts to promote variety in content, some readers perceive a bias in the material presented.

Apart from the need for more diverse opinions, voices and views, criticism also centres on the perceived aggression in the page's messaging. One of the participants, Zion, explains how she used to follow the page, but that what she experiences as 'aggression' in the posts drove her away from liking the page. However, she still follows it from a distance.

> **Zion:** I don't think the content is a problem. The problem is the way the content is shared. I feel as though the content is shared in a somewhat aggressive manner: I feel that they may be sharing a good message, but how the message is shared for me is what I feel has driven me from not liking the page again.

This extract clearly shows that women fans' selective consumption of the Zambian Feminists is closely tied to how they view the Administrator's roles. She is viewed as 'aggressive' in her approach, which leads participants like Zion to consume the Zambian Feminists content selectively. Interestingly, this aggressive approach is not enough to deter her altogether, for Zion still follows the page's discussions. One can then assume that the 32,000 plus followers of the Zambian Feminists page is not an accurate representation of follower numbers, as there may be other women fans who have since unliked the page but still avidly follow the discussions.

While participants complained that the Administrator's aggressive approach has driven some women away, they also acknowledge that it aligns with the page's tagline, 'of course we are angry', reflecting the frustration of Zambian women who feel unheard. This anger has polarized followers: if some are repelled by it, others feel it is necessary for advancing feminism in Zambia. Those who resonate with the aggression see it as a justified response to gender inequality. Just like feminist activists in the Global North, they want to use social media affordances to 'shout back' at the 'everyday sexism' of patriarchy and misogyny (Turley and Fisher, 2018: 130). Harriet exemplifies how women fans justify their anger and aggressive stance in feminism:

> **Harriet:** I resonate with the tagline. I am very angry because every time I wake up, all I see are stats about women and children being raped. I experience catcalling and sexism like every other day. In everything that I do I see how everything is about gender, how men are treated differently from women, how men earn more than women despite putting in the same work and how women do most of the work. So, I am angry about all these things, and I feel it is ok because I feel once you get angry, you get things done. Sometimes I feel you need to be more radical when it comes to dismantling these systems set up to oppress us.

Harriet's visceral response to her experiences of patriarchy serves to justify her anger and explain why her personal feelings resonate with the tagline of the Zambian Feminists page. Participants' perception of the Administrator as being aggressive in her approach is critical for understanding what passes for an acceptable public expression of anger among women and the expectation that women need to manage their anger. It provides a basis for understanding how women in patriarchal societies like Zambia have been socialized to view the public expression of anger as undesirable, and how transgressive it is to openly express feelings of outrage.

The interviews revealed that participants harbour deep anger towards the ongoing status quo of patriarchy, gender inequality and gender biases, yet they often refrain from expressing this anger publicly due to feelings of propriety associated with respectable femininity in the Zambian context. Despite criticizing the Zambian Feminists page for its perceived aggression, these participants admitted to feeling similar anger themselves but hesitated to express it openly. Chilombo, a university student, exemplifies this struggle:

> **Chilombo:** I am not a very confrontational person, I am quiet, and I try to be calm. However, inside, talking about feminist issues being told that "no, equality won't happen tomorrow" by men or women, which makes it even worse just boils inside me. I can't do anything about it because I feel like I will sound too angry, and I don't like that statement coming out of my mouth because we are expected to act a certain way by society. Women shouldn't be loud and angry, but we should be angry, and we have every right to be because we are being taken for a ride.

Chilombo is afraid of sounding 'too angry': her fear is one of feminine propriety in a cultural context which prohibits women expressing anger about feminist concerns. Her comment about letting the anger 'boil' inside her reaffirms Lerner's (1997) suggestion that women internalize social prohibitions against the expression of female anger. She feels a sense of powerlessness, unable to find a suitable outlet for her simmering frustrations. Nonetheless, Chilombo recognizes the importance of women's right to be angry and – in the interview at least – challenges societal norms that suppress such expressions. Social media and its affordances offer a channel for the expression of strong emotion, such as anger, and the social and individual parsing of 'affect' – the felt response to instances of gender injustice (Nau et al., 2023). We suggest that the Zambian Feminists page has become popular precisely because one of its functions is to openly express female anger in a space in which such is otherwise censured. By adding 'of course we are angry' to its tagline, the Zambian Feminists page gives Zambian women a platform and permission to express their anger and be heard publicly – if not personally, then at least vicariously.

Her page, but whose rules?

The Zambian Feminists Facebook page addresses contentious topics in Zambia, raising concerns about the safety of women contributors on this public platform. Public visibility allows internet trolls to potentially attack the Administrator and contributors. In an interview, the Administrator explained that she ensures safety by moderating comments:

> **Administrator:** I ensure safety by regulating comments. If someone posts something negative, hurtful, or demeaning, I instantly ban, block, and delete them. I don't want people to feel scared to talk about issues. I don't tolerate trolls and read all comments, banning as needed. Sometimes I use humour to tackle trolls; I clap and talkback, for example when a troll says something, I say "your mother". When I started, there were many trolls, but now there are almost none. It's my page, and I aim to create a safe space for open, fact-based conversations. While some may see this as unfair, we're serving a purpose by protecting minorities who need it.

The Administrator takes her page followers' safety seriously: she does not entertain trolls on her platform, and she bans, deletes and blocks anyone who posts anything she deems inappropriate. In this way she tries to ensure that the page is a safe space for women to share experiences and freely express themselves. She acknowledges that her approach may be viewed as one that is intolerant to different opinions: certainly, her no-nonsense attitude and actions appear to have been effective in deterring those who she judges to be provocateurs. Still, she justifies her actions by re-emphasizing the need to offer protection to minorities and the vulnerable on her page.

Another aspect of activism and online safety concerns the Administrator's personal safety. Some of the topics that she shares, such as advocating for LGBTQI rights in Zambia, are seen as uncultured, indecent or immoral. Zambia's laws do not permit homosexuality, putting her at risk of attack and arrest for her public advocacy. We sought to understand how she ensures her safety both online and offline.

> **Administrator:** I was once threatened with prison over a post about a manhunt for two girls kissing.[2] I thought it was stupid and spoke up against it, leading to calls for my arrest. I had to hide, take leave from work, and ask for the contact for Amnesty International. It was a risky move, but necessary. I also face personal attacks on my looks or weight. Building confidence is hard, but trolls can easily tear it down.

This brief anecdote reveals the different forms of intimidation Zambian feminist activists face, from personal attacks and character assassination by trolls to threats of imprisonment. This latter threat is taken seriously, for she felt obliged to turn to Amnesty International, an outside agency, for protection, a 'risky move': significantly, she does not mention a local equivalent for such an intervention.

What stood out in the Administrator's interview was her reference to Facebook itself as the biggest hindrance to her activism. Contradicting the received opinion that the internet is a free space where people can express themselves openly, the Facebook policy guidelines regarding what can be said on the platform potentially create a stumbling block for her activism:

> **Administrator:** Facebook drives me crazy with their restrictions. I've had more than ten posts banned for not meeting their standards, like saying "men are trash," which gets you a 48-hour ban. For example, I wrote about the "men are trash" movement, explaining it's not against all men but addresses the majority who commit crimes against women. When women stand up, the "not all men" argument surfaces. I compared this to South African xenophobic attacks, arguing that collective accountability can drive change, which is what feminists aim for. Despite a well-reasoned post, I got a 48-hour ban and was blocked from my page and personal account. Facebook's restrictions on what we can [do] drives me crazy.

This comment highlights Facebook as a double-edged sword, at times working against women's interests in favour of men. While community standards are meant to create a space for expression, phrases like 'men are trash' are instantly banned, whereas anti-feminist comments are allowed to remain (Curtis, 2018; Newton, 2019). While Facebook has admitted that the world is too diverse to take into consideration all the forms of hate speech (Zuylen-Wood, 2019), Facebook's existing policies pose challenges for African feminists trying to raise awareness or educate on women's issues.

Despite the challenge of Facebook's restrictions, the Zambian Feminists page offers women a platform to express their views. We asked our respondents if they feel safe expressing themselves on this platform considering Zambia's patriarchal context and the existence of trolls on Facebook. Interviews revealed that women feel supported when expressing their views on the feminist page. Nosiku, for example, claimed that she has not faced personal attacks for her opinions on the page. 'To be honest, I feel supported on the Zambian Feminist's page; I haven't gotten any opposing views, where someone comes to oppose me personally, I haven't gotten that.'

However, while women feel supported on the Zambian Feminists Facebook page, they prefer Twitter for advocacy. For them, Twitter offers a louder feminist voice and a more supportive community, allowing for freer expression without the presence of family and employers, unlike the more intolerant Facebook, confirming research in other spaces regarding Twitter's efficacy as a space for feminist advocacy (Turley and Fisher, 2018). They also attributed this to the Twitter demographic as one that is more open to different opinions and beliefs than Facebook:

> **Harriet:** On Facebook, I never felt supported, which is why I moved to Twitter. There, the feminist voice is louder, and people have diverse views. Twitter makes it easier to engage with feminists from Nigeria and Ghana compared to Facebook, where we must be friends to interact.

Harriet's comments highlight how Twitter's architecture and culture enhances the richness and diversity of feminist discussions. Twitter's default public setting allows for broader interaction, unlike Facebook, which requires connections to see posts (Twitter, 2021). This openness fosters global feminist connections on Twitter. This situation is not possible on Facebook because Facebook requires users to be connected in order to interact and see each other's posts (Boyd, 2007). While the Zambian Feminists page Administrator strives to create a safe space on Facebook, and the page provides support, many of the respondents appear to prefer Twitter for its diverse views and stronger feminist community.

Balancing dual identities: Online vs offline feminist identity

This section explores how the Zambian Feminists page Administrator and women fans negotiate the conflicting demands that arise between their online feminist identity and their offline roles as mothers, wives, sisters and daughters in a patriarchal society. Feminism in Zambia is contentious, and feminists are often seen as bitter and angry. It is perhaps not surprising then that the interviews reveal that some women maintain dual identities: boldly identifying as feminists online while rejecting the label in offline social settings in order to avoid attack or ostracism. This situation confirms Radloff's (2013) observation that women activists are met with a combination of online digital threats and offline issues that hinder them from publicly expressing, associating and actively participating as citizens. Nosiku reflects on this dilemma:

> **Interviewer:** Do you openly identify as a feminist?
> **Nosiku:** I do! Although I feel like a fraud sometimes. When asked if I'm a feminist, I've denied it despite knowing I am. In social settings where people mock feminists, I stay quiet to avoid confrontation. At home, I freely identify as a feminist, but online, I feel powerful and can argue all night.

Publicly identifying as a feminist in Zambia often invites push-back, so some, like Nosiku, reject the label to protect themselves from attacks. This duality makes them feel fraudulent, but it's a safety measure. Chilombo's comment further captures this struggle:

> **Chilombo:** It's easier to talk online than offline. I hate kneeling for men except my dad out of respect, which I tolerate as part of patriarchal culture. I am calmer with the men in my life, but I try not to negotiate too much. Online, I boldly proclaim "men are trash!" but reassure my boyfriend he's different. If someone asked me, "is your father also trash?" I would say yes. Though I acknowledge my father's role in patriarchy, I don't confront him directly to maintain our relationship.

Chilombo's comment shows how women must accommodate and balance their feminist beliefs with cultural practices and relationships. Women like Chilombo tread a fine line as they simultaneously reconcile themselves with observing cultural practices, such as kneeling to senior men as a sign of respect, try to protect and maintain important relationships with male family members and partners, and speak out against patriarchy and develop a feminist identity online. This situation highlights the fraught relationship between offline and online realities in contexts where identity is contested – at times violently. It illustrates Boyd's (2007) argument that the internet offers users the opportunity to forge new online identities, which can be multiple, or to reshape offline identity, carefully choosing 'what information to put forward, thereby eliminating visceral reactions that might have seeped out in everyday communication' (Boyd, 2007: 12). What Boyd's analysis is unable to show is the particularities of such negotiations and the social and personal costs that they may incur. Our women respondents also illustrate the argument made by Greijdanus et al. (2020) that people enact different personae online versus offline. Relatively anonymous online environments free people from concerns to be positively evaluated and consequent social restrictions to their behaviour (Greijdanus et al., 2020: 50). Despite this tension, some have integrated their identities, as Njavwa's evolution shows:

> **Njavwa:** Initially, I was an angry feminist online, avoiding the topic in non-feminist spaces to avoid ostracism. After a confrontation where I flipped a table, I realised I had to unify my identity. Now, everyone knows I'm a feminist, and I no longer hide it or fight with myself. I don't know if people are biting their tongues around me or they are really with my politics now, but whatever it is I am done with my closet.

Njavwa's comment highlights the importance of a unified identity and having the people around you know what you believe and stand for. Njavwa used to hide her feminism in certain social settings until she found herself in a situation where she had to defend her feminist values against a group of non-feminists. The resulting altercation resulted in her 'flipping' a table, an outward symbol of the frustration and anger she had been harbouring. Njavwa's journey emphasizes the importance of a unified identity, where friends and family respect her beliefs – or at least no longer confront her. While some participants are still working to integrate their identities, others, like Chanda, have managed to achieve a coherent character:

> **Chanda:** The Chanda you see online is the same everywhere. I'm a strong feminist both with my kids and at work I am a strong woman everywhere.

Chanda's consistency suggests that maintaining a single feminist identity online and offline is possible, even in Zambia.

Conclusion

Zambian women are confronted with a highly patriarchal social order, established by a meshing together of colonial, customary and Christian practices and values. Unsurprisingly, they must struggle to make their voices heard. Feminism, such as that found on the Zambian Feminists Facebook page, challenges local patriarchal limitations, promoting flexibility in gender definitions and enhancing democratic citizenship by opposing exclusionary practices and fostering new spaces for belonging and association. Following Nyamnjoh's (2006) idea that citizenship is an ongoing process rather than a final goal, we view Zambian feminism online as a complex and evolving confluence of identities and practices that must continuously work – albeit somewhat unevenly – towards gender justice. The Zambian Feminists page supports an online community that shares experiences and provides encouragement, both online and offline, to empower women followers to challenge oppressive gender norms, as Cochrane (2013) and Munro (2013) suggest. The 'anger' of the tagline is not only central to the page's identity, but it also becomes a means to decipher the page's popularity as well as its expressive social function for its followers. Rather than dismissing it as merely a feminine outburst, we argue that it calls attention to the structural challenges faced by women who are otherwise obliged to conform to the respectable self-control demanded of them within Christian and customary patriarchy. Supported by their proactive Administrator, who manages and deflects the opprobrium generated by the page's outspoken stance, the page provides Zambian women with a space in which to safely challenge oppressive gender norms while they manage the banal yet persistent demands of patriarchy. In effect, the page is used by its followers as a testing ground for the limits of feminist practice.

That these women live in Lusaka, with access to urban amenities including the technological infrastructures, data and devices needed in order to be active on social media, is of consequence: these findings are not necessarily applicable to all Zambian women. In this regard, and addressing Roberts and Bosch's (2023: 20) question about the conditions needed to use digital technology for social change, we observe that in Lusaka various discourses on gender roles – customary, Christian and Western – compete, shaping perceptions of acceptable male and female behaviour. These contesting ideas create a unique context in which followers of the Zambian Feminists Facebook page develop their feminist identities and seek more just and equitable forms of social and cultural belonging, or citizenship. Given Zambia's gender-unequal context, it is unsurprising that participants struggle to maintain both online and offline feminist identities. Our findings indicate that these identities can be mutually exclusive or complementary, driven by fear of social ostracism or surveillance. This dual identity serves as a safety measure in a society where feminism remains contentious: rather than an instance of feminist failure, it is a timely reminder of the flexibility required to mitigate and navigate the social costs of feminist practice.

Notes

1. Former US Ambassador to Zambia, Daniel Foote, publicly condemned the fifteen-year prison sentence for a gay couple for behaviour that President Edgar Lungu defended as 'unbiblical and unchristian'. Foote was recalled, and the incident sparked widespread debate on social media in Zambia, where conflicting and unpopular opinions were contested (Foote, 2019).
2. On 30 January 2018, police launched a search for a lesbian couple whose pictures went viral on social media alongside a story purporting them to be in an intimate relationship (Chabala, 2018).

Bibliography

Allen, J. (2010), 'Gender, British Administration and Mission Management of Education in Zambia 1900-1939', *Journal of Education Administration and History*, 42 (2): 181–92.

Ault, J. M. (1983), 'Making "Modern" Marriage "Traditional": State Power and the Regulation of Marriage in Colonial Zambia', *Theory and Society*, 12(2): 181–210.

Becker, H. S. and Geer, B. (1957), 'Participant Observation and Interviewing: A Comparison', *Human Organization*, 16, 28–32.

Boyd, D. M. (2007) 'Why Youth (Heart) Social Network Sites: The Role of Networked Publics in Teenage Social Life, in D. Buckingham (ed.), *MacArthur Foundation Series on Digital Learning- Youth, Identity, and Digital Media Volume*, Cambridge: MIT Press.

Bryman, A. (2012), *Social Research Methods*, 4th edn, New York: Oxford University Press.

Chabala, M. (2018), 'Police Hunt for Lesbian Couple', *News Diggers Online*, 30 January. Available at: https://diggers.news/local/2018/01/30/police-hunt-for-lesbian-couple/ (Accessed: 8 July 2024).

Cochrane, K. (2013), 'The Fourth Wave of Feminism: Meet the Rebel Women', *The Guardian Online*, 10 December. Available at: https://www.theguardian.com/world/2013/dec/10/fourth-wave-feminism-rebel-women (Accessed: 8 July 2024).

Connell, R. W. (2009), *Gender in World Perspective*, 2nd edn, Cambridge: Polity Press.

Crehan, K. (1997), *The Fractured Community: Landscapes of Power and Gender in Rural Zambia*, Berkeley: University of California Press.

Curtis, C. (2018), 'Facebook Thinks Saying "All Men are Trash" is Hate Speech — it's Not', *TNW Online*, 6 September. Available at: https://thenextweb.com/facebook/2018/09/06/facebooks-double-standards-censorship/ (Accessed: 8 July 2024).

DataReportal (2023), *Digital 2023: Zambia*. Available at: https://datareportal.com/reports/digital-2023-zambia?rq=zambia (Accessed: 8 July 2024).

Deacon, D., Pickering, M., Golding, P. and Murdock, G. (1999), *Researching Communications: A Practical Guide to Methods in Media and Cultural Analysis*, London: Arnold.

Epstein, A. L. (1981), *Urbanisation and Kinship: The Domestic Domain on the Copperbelt of Zambia, 1950–1956*, London: Academic Press.

Evans, A. (2014), '"Women Can Do What Men Can Do": The Causes and Consequences of Growing Flexibility in Gender Divisions of Labour in Kitwe, Zambia', *Journal of Southern African Studies*, 40 (5): 981–98.

Evans, A. (2015), 'History Lessons for Gender Equality from the Zambian Copperbelt, 1900–1990', *Gender, Place & Culture*, 22 (3): 344–62.

Evans, A. (2018), 'Cities as Catalysts of Gendered Social Change? Reflections from Zambia', *Annals of the American Association of Geographers*, 108 (4): 1096–114.

Evans-Pritchard, E. E. (1981), *A History of Anthropological Thought*, London: Faber and Faber.

Ferguson, J. (1999), *Expectations of Modernity: Myths and Meanings of Urban Life on the Zambian Copperbelt*, London: University of California Press.

Foote, D. (2019), 'Press Statement from U.S. Ambassador Daniel L. Foote', *U.S. Embassy in Zambia*, 2 December. Available at: https://zm.usembassy.gov/press-statement-from-u-s-ambassador-daniel-l-foote/ (Accessed: 8 July 2024).

Gill, J. and Johnson, P. (2002), *Research Methods for Managers*, 3rd edn. London: Sage Publishing.

Greijdanus, H., Fernandes, C., Turner-Zwinkels, F., Honari, A., Roos, C., Rosenbusch, H. and Postmes, T. (2020), 'The Psychology of Online Activism and Social Movements: Relations Between Online and Offline Collective Action', *Current Opinion in Psychology*, 35, 49–54.

Jensen, K. B. (1988), 'Answering the Question: Why Reception Analysis?', *Nordic Review*, 9 (1): 3–5.

Lerner, H. (1997), *The Dance of Anger*, New York: Harper Collins.

Lunt, P. and Livingstone, S. (1996), 'Rethinking the Focus Group in Media and Communications Research', *Journal of Communication*, 46 (2): 79–98.

Lyster, R. (2019), 'The Death of Uyinene Mrwetyana and the Rise of South Africa's "Am I Next?" Movement', *The New Yorker*, 12 September. Available at: https://www.newyorker.com/news/news-desk/the-death-of-uyinene-mrwetyana-and-the-rise-of-south-africas-aminext-movement (Accessed: 8 July 2024).

Munro, E. (2013), 'Feminism: A Fourth Wave?', *Political Insight*, 4 (2): 22–5.

Nau, C., Zhang, J., Quan-Haase A. and Mendes, K. (2023), 'Vernacular Practices in Digital Feminist Activism on Twitter: Deconstructing Affect and Emotion in the #MeToo Movement', *Feminist Media Studies*, 23 (5): 2046–62, DOI: 10.1080/14680777.2022.2027496.

Newton, C. (2019), 'Why You Can't Say "Men Are Trash" on Facebook', *The Verge Online*, 3 October. Available at: https://www.theverge.com/interface/2019/10/3/20895119/facebook-men-are-trash-hate-speech-zuckerberg-leaked-audio (Accessed: 8 July 2024).

Nothias, T. (2020), 'Access Granted: Facebook's Free Basics in Africa', *Media, Culture & Society*, 42 (3): 329–48.

Nyamnjoh, F. (2006), *Insiders and Outsiders: Citizenship and Xenophobia in Contemporary Southern Africa*, New York: Zed Books.

Parpart, J. (1991), *Gender, Ideology and Power: Marriage in the Colonial Copperbelt towns of Zambia*. Available at: https://wiredspace.wits.ac.za/items/c635dcf3-55a5-4d60-8f2a-2b7202c60ef7 (Accessed: 8 July 2024).

Parpart, J. (1994), '"Where is Your Mother?": Gender, Urban Marriage, and Colonial Discourse on the Zambian Copperbelt, 1924-1945', *The International Journal of African Historical Studies*, 27 (2): 241–71.

Parpart, J. (1986), 'Class and Gender on the Copperbelt: Women in Northern Rhodesian Copper Mining Communities, 1926–1964', in C. Robertson and I. Berger (eds), *Women and Class in Africa*, New York: Africana Publishing Company.

Rasing, T. (2001), *The Bush Burnt, the Stones Remain: Female Initiation Rites in Urban Zambia*, Netherlands: LIT Verlag.

Radloff, J. (2013), Digital Security as Feminist Practice, *Feminist Africa*, 145–55.

Ray, R. (2018), 'Postcoloniality and the Sociology of Gender', in J. Messerschmidt, P. Yancey Martin, M. Messner and R. Connell (eds), *Gender Reckonings: New Social Theory and Research*, New York: New York University Press, 73–89.

Richards, A. (1940), *Bemba Marriage and Present Economic Conditions*, Livingstone: The Rhodes-Livingstone Institute (The Rhodes-Livingstone Paper number 4).

Roberts, T. and Bosch, T. (2023), 'Introduction: Spaces of Digital Citizenship in Africa', in T. Roberts and T. Bosch (eds), *Digital Citizenship in Africa: Technologies of Agency and Repression*, London: Zed Books, 1–32.

Schrøder, K., Drotner, K., Kline, S. and Murray, C. (2003), *Researching Audiences*, London: Hodder Arnold.

Siachitema, K. (2010), 'Experiences of Gender Mainstreaming in Zambia', in T. Matebu, and D. Abiye (eds), *Gender Mainstreaming Experiences in Eastern and Southern Africa*, Addis Ababa: OSSREA, 78–92.

Siwila, L. (2017), 'Reconstructing the Distorted Image of Women as Reproductive Labour on the Copperbelt Mines in Zambia (1920-1954)', *Journal for the Study of Religion*, 30 (2): 75–89.

Solon, O. (2017), '"It's Digital Colonialism": How Facebook's Free Internet Service has Failed its Users', *The Guardian*, 27 July. Available at: https://www.theguardian.com/technology/2017/jul/27/facebook-free-basics-developing-markets (Accessed: 8 July 2024).

Taylor, S. (2006), *Culture and Customs of Zambia*, London: Greenwood Press.

Turley, E. and Fisher, J. (2018), 'Tweeting Back While Shouting Back: Social Media and Feminist Activism', *Feminism & Psychology*, 28 (1): 128–32.

Twitter (2021), *About Public and Protected Tweets*. Available at: https://help.twitter.com/en/safety-and-security/public-and-protected-tweets (Accessed: 10 July 2022).

Willems, W. (2013), 'Participation – In What? Radio, Convergence and the Corporate Logic of Audience Input Through New Media in Zambia', *Telematics and Informatics*, 30 (3): 223–31.

Willems, W. (2016), 'Beyond Free Basics: Facebook, Data Bundles and Zambia's Social Media Internet', *LSE Africa Blog*, 1 September. Available at: https://blogs.lse.ac.uk/africaatlse/2016/09/01/beyond-free-basics-facebook-data-bundles-and-zambias-social-media-internet/ (Accessed: 10 July 2022).

World Bank (2021), *Individuals Using the Internet (% of Population) - Zambia*. Available at: https://data.worldbank.org/indicator/IT.NET.USER.ZS?locations=ZM (Accessed: 8 July 2024).

Zuylen-Wood, S. (2019), '"Men Are Scum": Inside Facebook's War on Hate Speech', *Vanity Fair Online*, February 26. Retrieved 8 July 2024 from https://www.vanityfair.com/news/2019/02/men-are-scum-inside-facebook-war-on-hate-speech.

Chapter 4

KEEPING EACH OTHER SAFE: TRANSGENDER PEOPLE'S ONLINE SOLIDARITY STRATEGIES

Nyx McLean

Transgender, non-binary and gender-diverse (TNBGD) people are often subjected to violence based on their gender identity; this is inclusive of technology-facilitated violence. This chapter is concerned with the experiences of TNBGD people in Botswana, Rwanda, South Africa and Uganda and the nature of the technology-facilitated violence they are subjected to due to their gender identity and expression, and names this violence as online gender-based violence (OGBV). Through a feminist internet research design, the chapter seeks to understand the ways that TNBGD people keep each other safe online and build counterpower to technology-facilitated violence. Feminist internet research extends to the online and digital technologies the work of feminist research of interrogating and disrupting power, and dismantling systemic oppression to create a more just and inclusive reality (Hesse-Biber 2007; Cooky, Linabary and Corple, 2018; Wigginton and LaFrance, 2019; hooks, 2000).

The online solidarity strategies that TNBGD people draw on to navigate and manage their safety online form a distinctive aspect of feminist digital citizenship that this chapter seeks to highlight. TNBGD people reported mobilizing with others to keep especially vulnerable individuals safe when targeted by transphobic individuals and groups. This chapter argues that this tactic is a strategic response and resistance to the absence of effective reporting mechanisms and adequate responses from social media platforms, and that, in doing so, TNBGD people are actively building counterpower and participating in feminist digital citizenship. This chapter presents a contextual overview of transphobia in Botswana, Rwanda, South Africa and Uganda before presenting a review of literature on online safe spaces, TNBGD experiences of OGBV, and organizing online, and then going on to present the feminist internet research design. The analysis in this chapter centres on how TNBGD people navigate, manage and stay safe online and focuses the discussion on the safety mechanisms used to do so, such as closed groups, blocking and reporting, and solidarity strategies. The chapter concludes with recommendations made for both TNBGD people and social media platforms in order to ensure the online safety and continued participation of TNBGD people.

Background

This chapter is written from the findings of a broader project, The Left Out Project, which sought to explore TNBGD people's experiences of OGBV in Botswana, Rwanda, South Africa and Uganda (McLean and Cicero, 2023). The project centred TNBGD people's experiences of OGBV so that responses, such as policies and interventions, to OGBV may be better informed and ensure stronger protections for TNBGD people.

Transphobic violence manifests in a number of ways, which can range from verbal to physical abuse, sexual violence and even murder. Transgender women, in particular, experience the highest rates of violence (Muller et al., 2021; Graaff, 2021). In 2021, Muller et al., in their nine African countries' study, reported that, 'three in four transgender women (73 per cent) had experienced a form of violence in their lifetime, and almost half (45 per cent) in the past year' (Muller et al., 2021: 4). Transphobia and the threat of transphobic violence have a significant impact on TNBGD people's mental health (Lanham et al., 2019).

Violence against LGBTQIA+ people, and particularly TNBGD people, has increased worldwide (McLean and Cicero, 2023; Human Rights Campaign, 2023). The spike in hate crime-related murders is a result of a 'culture of violence' that emerges from transphobia and intersecting discriminations on the basis of racism, sexism and homophobia (Human Rights Campaign, 2020). On the African continent, LGBTQIA+ rights are in a dismal state, and as Iranti (2019: 5) reports, 'violence against LGBTQI+ people is a critical area of concern for the African continent'. Where countries do afford some state protections, such as in the case of Botswana and South Africa, these protections do not always translate from paper to people, in that high rates of homophobic and transphobic violence are still found (McLean, 2020; Mhaka, 2021). For instance, in South Africa, Gender DynamiX (2021), 'documented 60 cases of human rights violations perpetrated against trans and gender diverse persons in the space of three months'.

The Left Out Project found that the violence affecting TNBGD people is widespread across Botswana, Rwanda, South Africa and Uganda, and largely driven by systemic, cultural, religious and political influence. Participants from all four countries reported discrimination from healthcare professionals, experiences of being arrested and detained, sexual harassment and violence, online harassment and being targeted by transphobic groups, to name only a few (see, for instance, Igonya, 2022; Nyeko, 2022; Andresen, 2020; Epprecht, 2019; McLean, 2018; Reid, 2017; McLean, 2015; Wells and Polders, 2006; Teal and Conover-Williams, 2016; Thoreson, 2008).

This study found that TNBGD people in Botswana, Rwanda, South Africa and Uganda encounter frequent and consistent technology-facilitated violence that targets their gender identity, thus being defined as OGBV. OGBV is understood as gender-based violence that is enacted and enabled through the use of digital technologies and includes infringement of privacy, surveillance and monitoring, damaging one's reputation and/or credibility, harassment, direct threats and/ or violence, and targeted attacks to communities (Association for Progressive

Communications, 2017: 4-6). The TNBGD participants in this study felt that their experiences should be considered to be OGBV given the nature of the attacks directed towards them.

A non-binary person from Botswana (B4) explained that for them, the violence they were on the receiving end of constitutes OGBV because of 'the way the violence specifically attacks my gender identity, I consider that as gender-based violence'. This was reinforced by a non-binary trans femme person from South Africa (S1) who shared that 'if you just look at the semantic definition of gender-based violence, it is violence directed at you, because of your gender. People are being violent towards me because I'm transgender, and that is my gender identity. Therefore, it is online gender-based violence'.

The above two examples do make it seem clear that the violence experienced by TNBGD people should be considered to be OGBV. However, current conceptualizations are oversimplified as meaning violence against women, specifically cisgender women. While OGBV is defined as violence that is experienced as a direct result of one's gender identity and expression, conceptualizations do not include a diversity of gender identities in their responses to OGBV. If OGBV responses are not inclusive of all gender identities and expressions, TNBGD people are at risk of continued discrimination and violence, and social reality continues to be misrepresented (Jauk, 2013).

Participants shared some of their experiences of OGBV, speaking first to the relentlessness of the OGBV they experience. For instance, a trans woman in Uganda (U3) shared how, 'every time I open most of my social media apps, you will find an insult … it's a daily thing'. While a trans woman in Rwanda shared, 'there are [so] many times that I cannot count them'. A non-binary South African (S5) said that for them, 'I can't remember a time when there wasn't transphobia online'.

These three examples show the continuous nature of the OGBV targeting TNBGD people that they must navigate online. OGBV significantly impacts TNBGD people's digital citizenship through the threat or real presence of violence, which may result in them self-censoring or outright leaving social media platforms in order to remain safe. The navigation of this violence and the counterpower strategies employed by TNBGD people as resistance to the OGBV they experience are the focus of this chapter.

Literature review

Transgender communities are some of the most vulnerable groups globally and the target of hate-based violence, even in countries that afford legal protections to LGBTQIA+ people (Lombardi et al., 2002). Transphobia, as defined by the Triangle Project, is 'an irrational fear of and/or hostility towards people who are transgendered or who otherwise transgress traditional gender norms' (Lee, Lynch and Clayton, 2013: 6). Transphobic harassment and violence may manifest as verbal abuse, physical or sexual violence and even murder. Transgender women are the most vulnerable to transphobic violence, as reported in a study conducted by Muller et al. (2021: 4), with 73 per cent of transgender women experiencing some form of transphobic violence in their lifetime. The consequences of the threat of transphobic violence are significant, in particular in relation to transgender people's mental health (Lanham et al., 2019).

TNBGD people may be further marginalized by poverty (McLean, 2020; McLean, 2018; Thoreson, 2008). Transphobic discrimination is amplified by socio-economic inequality and poverty whereby TNBGD people may not have the financial capital to exist independently of their families. Further, it is important to note the frequent occurrence in some African countries where religious leaders hold significant sway, and often attribute drought, poverty, a pandemic like Covid-19 and any other hardship a community endures as 'divine punishment' for harbouring homosexuals (Reid, 2020). This makes LGBTQIA+ people targets of harassment and violence in their communities. Violence against LGBTQIA+ people, and particularly TNBGD people, has increased worldwide (Human Rights Campaign, 2023). It bears noting that police reports and crime statistics do not accurately reflect the degree of transphobic crime due to the misgendering of transgender people by police, the criminal justice system and the media (Graaff, 2021).

Safer spaces online

The internet and digital technologies are often understood to be safer spaces for LGBTQIA+ people, such as many of those who live on the African continent, where offline spaces may be violent and dangerous to these groups (McLean, 2020; McLean and Mugo, 2015; Schudson and Van Anders, 2019). Digital spaces can be negotiated, navigated and managed – such as through closed groups and secure/encrypted communication (McLean, 2020; McLean and Mugo, 2015). Some members of these online safe spaces may form connections and migrate to text-based messaging applications, such as WhatsApp, or into offline spaces. However, sharing of contact data and/or physical space details may not be secure, given privacy controversies with some social media platforms and their relationships with state governments (McLean, 2020; Article 19, 2015; McLean and Mugo, 2015).

Previous research has shown how broadly on the African continent, and more specifically in South Africa, the internet makes it possible for LGBTQIA+ people to practice digital citizenship, to connect with each other and to form safe digital spaces in contexts where forming safe spaces offline may be impossible (McLean, 2020; McLean and Mugo, 2015). However, it also provides people, communities and governments with tools to monitor LGBTQIA+ people, especially in African countries with anti-homosexuality laws (McLean, 2020). As Nigro (2019) writes, LGBTQIA+ people as a group, 'long criminalized and pathologized, is often an afterthought when it comes to user privacy and regulations – which has resulted in a precarious digital landscape'. The reason for this being that the context in which technology is built is a neoliberal one that does not consider how technologies may exacerbate oppressions, such as those founded in race, gender, class and sexuality (Noble, 2018: 1). Or, in the case of Facebook, it actively excludes TNBGD people through policies such as the 'real names' policy (Haimson and Hoffmann, 2016; Article 19, 2015). Dating apps, for instance, often make use of location data to link users to each other but, while useful in finding a match, this can also put users at risk of harassment and violence from those making use of the apps to target LGBTQIA+ people. For instance, dating apps by using location data may make it easier to identify and target LGBTQIA+ people (Van Zyl and McLean, 2021). Van Zyl and McLean (2021) provide the example of an Egyptian gay man who was located by police through the gay dating app Grindr and arrested. He had not met with anyone offline, but the argument was presented that his use of the app was evidence enough to arrest him (Van Zyl and McLean, 2021). It is important to protect the data and privacy of users, especially for LGBTQIA+ people, who are at risk of discrimination and persecution if their identities are made known to homophobic, transphobic and queerphobic communities and state governments.

TNBGD experiences of OGBV

OGBV is violence that is facilitated through the use of technology, and it significantly impacts the ability of individuals to participate in digital spaces (Henry, Vasil and Witt, 2022; Henry and Powell, 2018). OGBV is defined as 'acts of gender-based violence that are committed, abetted or aggravated, in part or fully, by the use of information and communication technologies (ICTs), such as mobile phones, the Internet, social media platforms, and email' (Association for Progressive Communications, 2018; see also Suzor et al., 2018; Faith, 2018). OGBV includes infringement of privacy, surveillance and monitoring, damaging reputation and/or credibility, harassment (which may be accompanied by offline harassment), direct threats and/or violence and targeted attacks on communities (Association for Progressive Communications, 2017: 4-6). It also bears noting that OGBV and offline GBV are often entangled with each other, where OGBV may lead to offline GBV, or offline GBV may occur in conjunction with OGBV (see, for instance, Web Foundation, 2020).

Current conceptualizations of OGBV neglect and invisibilize the experiences of TNBGD people by oversimplifying OGBV as meaning violence against women – specifically, cisgender women (Jauk, 2013: 808-9). Not only then are current conceptualizations oversimplified, but they are also informed by cisgenderism, which sees gender as a binary of cisgender women and men and does not recognize or include other gender identities, such as TNBGD identities (Rogers, 2021: 2190). This cisgenderist lens renders TNBGD experiences invisible. Because these identities are not recognized as real or valid, they are 'neglect[ed] ... within research, policy, and practice' (Rogers, 2021: 2201-2). The consequences of this are that resources, support and interventions are not made available for TNBGD people experiencing OGBV. For instance, information materials may rely on binary gendered language and visuals that 'can reinforce perceptions that such resources are reserved for cisgender (not transgender) women' (Jordan, Mehrotra and Fujikawa, 2020: 541).

If TNBGD identities are not recognized or protected in a country, crime statistics may be skewed. For example, violence against trans women may not be documented appropriately and included in GBV crime statistics because the victims have been misgendered as cisgender men (Graaff, 2021: 6-7). This narrow definition of GBV and gender identity is deeply harmful, especially considering that 'across all intersecting identities, trans women are reported to experience higher rates of violence than almost any other group' (Graaff, 2021: 7). This is a key group that requires social, legal and other protective measures but that is denied this right because current OGBV frameworks are too narrow and are exclusionary. However, it does require noting that the expansion of the OGBV definition to include TNBGD identities must be careful not to perpetuate another binary – the transgender binary of trans men and trans women. Non-binary and gender-diverse people are often not included in conversations about transgender identities, but it is critical that they are included – especially considering that this group makes up a third of the overall TNBGD community (Matsuno et al., 2022).

Studies exist on the use of the internet and digital technologies by marginalized groups, but very few have studied these groups' experience of online violence, let alone OGBV. Powell, Scott and Henry (2020) provide one of the very few discussions on digital harassment experienced by LGBTQIA+ people. Their study examined the experiences of Australian and British adults (aged 18 to 54). A key finding of their study was that 'transgender individuals experience higher rates of digital harassment and abuse overall, and higher rates of sexual, sexuality and gender based harassment and abuse, as compared with heterosexual cisgender individuals' (Powell, Scott and Henry, 2020: 199). Globally, there have been several reports, such as those from the Human Rights Campaign (2023), examining violence against TNBGD people. The Human Rights Campaign primarily monitors violence against US transgender people, but they do not, however, report on OGBV against TNBGD people. This is important to note because OGBV may lead to targeted attacks offline, potentially resulting in the murder of TNBGD people. Given the severity of the risk to TNBGD people, there is a need for an inclusive understanding of GBV and OGBV to ensure that TNBGD experiences are accounted for in responses to OGBV (Graaff, 2021; Lombardi et al., 2002). TNBGD people are at risk of continued discrimination and violence if such an expansion does not take place.

Organizing online: Feminist digital citizenship

Digital technologies, such as the internet, social media platforms and mobile phones, have enabled a greater sense of connectedness among people (Roberts and Bosch, 2023; Henry, Vasil and Witt, 2022). Digital citizenship is understood as the uptake of digital technologies by citizens in order to actively participate in their communities and society. This includes the political use of digital technologies to advocate for their rights and resist violence and power through gathering and sharing information, connecting with community members, organizing and driving for social change (Roberts and Bosch, 2023: 2). This is particularly significant for vulnerable groups such as TNBGD people, who may feel that they have little agency when facing social and often government-sanctioned prejudice and discrimination, including the risk of violence and threat to life (McLean and Cicero, 2023; McLean, 2020). It is through the use of digital technologies that they may actively resist such structures and advocate for their rights.

While digital technologies have enabled citizens to participate more fully online, advocate for social change and build a greater sense of connectedness and community with each other, they may also be used to enable mediated violence, such as hateful targeting, hacking, stalking and surveillance of TNBGD people (McLean and Cicero, 2023; Roberts and Bosch, 2023). This affects the scope and space for active digital citizen engagement with issues critical to their well-being (Roberts and Bosch, 2023; McLean and Cicero, 2023). For instance, TNBGD people may opt out of engaging on social media platforms to keep themselves safe and avoid violent interactions (McLean and Cicero, 2023). This withdrawal and self-censorship lead to what is known as the 'chilling effect' (Posetti et al., 2022: 12). Safe access to and use of digital technology is a critical matter of consideration when speaking to digital citizenship because digital safety and security determine to what degree TNBGD people, as well as other vulnerable groups, may feel able to actively participate online (McLean and Cicero, 2023; Ajaja, 2023).

Feminist digital citizenship is concerned with how individuals, in particular women and TNBGD people and other marginalized groups, take up digital technologies to participate online, organize and advocate for social change, with an emphasis on agency, rights, activism and other acts of resistance. Gender and associated power relations, including inequality and experiences of violence, are rarely accounted for in conceptualizations of digital citizenship (Henry, Vasil and Witt, 2022: 1973). A feminist reading of digital citizenship is necessary to account for how gender, power and inequalities inform experiences and practice of digital citizenship – particularly when considering the manner by which activists have taken up the use of digital technologies and online spaces 'as a crucial part of their activism', especially considering traditional media, which ignore or silence the voices of activists and marginalized groups (Mudavanhu and Radloff, 2013: 328). Digital technology 'is now more likely than ever to accompany face-to-face feminist organising' (Pereira, 2017: 21). The use of digital technology, however, is not passive. Rather, digital feminist activists actively shape digital media content in ways that are resistant to subjugating power and inequality (Yun, 2020: 61). This places emphasis on the agency and active uptake of digital technology by feminist activists in shaping their reality and grappling with issues of power.

Feminist activism online is political participation, and this is feminist citizenship in practice. Yun (2020: 59) argues that what matters online is not geographic proximity but rather 'the intensity of digital connections based on networked affect'. Feminists make use of digital technology, such as social media platforms, to remain connected, to participate and organize action online and offline, 'enabl[ing] feminist solidarity across space and time' (Yun, 2020: 60). It is this sense of connectedness and solidarity that is of interest in this chapter.

Methodology and research design

In seeking to understand TNBGD people's experiences of OGBV, their navigation thereof and the strategies they employ to keep themselves safe, this chapter draws on a qualitative feminist internet research approach primarily influenced by standpoint theory, intersectionality, reflexivity and feminist ethics of care (McLean, 2022). Interviews were conducted with TNBGD participants from Botswana, Rwanda, South Africa and Uganda to explore their experiences, and this data was then analysed using thematic analysis.

Qualitative research was deemed a better fit for the study, given that as an approach it makes it possible to gather in-depth and meaningful knowledge that enables exploration of the experiences of TNBGD people with regards to OGBV (Staples et al., 2018). In seeking to understand digital experiences informed by gender identity and expression, feminist internet research proved to be an appropriate fit for the study. The feminist internet research methodological and ethical framework employed in this study is one which is upheld by four key pillars, namely, standpoint theory, intersectionality, reflexivity and feminist ethics of care (McLean, 2022). The pillars are discussed very briefly below.

Standpoint theory places emphasis on the lived experiences of those who are marginalized and holds that 'knowledge is always socially situated' (Harding, 2004). Standpoint theory is not only about knowing the conditions of oppressed groups but also about critically engaging with these positions, eliciting key insights and using this knowledge and understanding for the purposes of emancipation and knowledge building (Fawcett and Hearn, 2004). Standpoint theory informed the research on TNBGD experiences through centring the wealth of knowledge of TNBGD people's lived experiences and emphasizing that they are better informed than anyone else on their experiences.

Intersectionality is the understanding that all forms of struggle and oppression are linked and overlap through race, gender identities, nationality, religion and sexuality (Crenshaw, 1989, 1991; McLean, 2022). Crenshaw introduced the concept of intersectionality to show how discriminations based on gender and race, and other identity-based discriminations, exacerbate each other and cannot be separated (Crenshaw, 1989). This helps to illustrate that participants may experience multiple forms of discrimination that may be compounded and result in a unique experience of marginalization. The key identity aspect that was identified as intersecting with gender identity was sexual identity as people who identify as part of the LGBTQIA+ community. This is, unfortunately, not a surprising overlap and exacerbation of discrimination given the current climate of push-back on LGBTQIA+ rights and increase in hate-based anti-LGBTQIA+ rhetoric.

Reflexivity makes it possible for researchers to engage with their own positionality, power and privilege (Hundle, Szeman and Hoare, 2019; Rodríguez-Dorans, 2018; McLean, 2022). It asks that researchers critically consider their positionality and how this impacts on the research process and their participants. Reflexivity allows researchers to engage with how their social location, privilege, values and assumptions, to name a few, may influence the research process (Hundle, Szeman and Hoare, 2019). For instance, the author of this chapter identifies as a trans non-binary person, and this positionality, in addition to other identity experiences, influenced their experiences and interactions with the participants.

Feminist ethics of care is centred on the concern for the research community and participants and asks researchers to consider whether their work benefits participants and communities or is extractive and perpetuates injustice (Preissle, 2007; Edwards and Mauthner, 2012; Knapp and Ogunbanjo, 2009). Incorporating ethics of care into the research makes provision for considerations regarding sensitivities surrounding OGBV and the violent experiences participants may have encountered. Using feminist ethics of care, it was possible to create an environment of care that responded to the needs of participants. For instance, LGBTQIA+ activists, organizations and networks were contacted in order to facilitate access to participants and to ensure that support structures for participants were in place when needed.

It does need to be noted that given the state of LGBTQIA+ rights on the African continent, this is a particularly vulnerable group, and their safety was of the utmost importance. To facilitate this process of keeping the identities of the participants safe, especially in areas where TNBGD individuals are not protected, participants remained anonymous, and in Rwanda and Uganda, in particular, digital security measures such as the use of encrypted email accounts supported by Riseup and virtual private networks (VPNs) were used for communication.

Throughout the research process, especially during data collection, analysis and dissemination, the four pillars of feminist internet research (McLean, 2022) were consistently drawn on. This was to ensure that the study accounted for the lived experiences of the TNBGD participants, took into consideration intersecting discriminations and how these intersected with the author's lived experiences and lens, and assessed the potential risks and harms that participating in such a study could have on the individuals and their communities involved in this study.

The research was conducted in September and October 2022 with twenty-nine adult individuals aged 18 and older who resided in Botswana, Rwanda, South Africa and Uganda. The study included two countries where LGBTQIA+ rights are legalized, namely, Botswana and South Africa, and two countries where LGBTQIA+ rights are not protected, these countries being Rwanda and Uganda. The reasoning behind this was to explore how TNBGD people's experiences of OGBV would differ or if common themes would emerge across some or all of the countries.

Given the vulnerability of TNBGD people, letter and number identifiers were used to address concerns around the possibility of participants being identifiable. For Botswana, the letter B, Rwanda, the letter R, South Africa, the letter S, and Uganda, the letter U. The numbering assigned to participants is random.

The study made use of in-depth qualitative interviews, which were conducted online. Online interviews were conducted on a one-on-one basis through remote video recording technologies. Connectivity was not always good in all the countries, and where connectivity failed outright on a first and a second attempt to conduct the interview, the remaining questions were sent to participants via email so that they could respond via text. The data was analysed using thematic analysis (Guest, MacQueen and Namey, 2012, Ponnam and Dawra, 2013; Clarke, Braun and Hayfield, 2006). Thematic analysis involves the identification, analysis and interpretation of patterns or themes that exist within the data. Steps followed in a thematic analysis include: 1) becoming familiar with the data, for instance; 2) coding or searching for themes; 3) reviewing themes; 4) refining themes; and lastly 5) writing up the findings, which is where the themes are drawn together, analysed and discussed (Clarke, Braun and Hayfield, 2006). Once these steps have been followed, the themes and related data are brought into conversation with literature to deepen the analysis. It bears noting that the themes that emerged from the data cut across all country contexts and have thus been presented collectively.

Analysis

TNBGD people make use of digital technologies, such as social media, in affirming ways in order to celebrate their gender identity, and this may include taking and sharing photographs of themselves in clothing that matches their gender identity. It is this content that receives targeted abuse online. Online acts of violence can be reflective of conservative ideologies as well as cultural beliefs and norms around gender identity, sexual identity and expression of both gender and sexual identity, even in countries where such expression is not restricted through censorship laws (Gupta, 2019). Acts of violence may have a silencing and policing effect on TNBGD people persons, which is consistent with what the TNBGD participants reported in this study. The discussion below presents some of the experiences of OGBV and then turns to the heart of this chapter: the strategies employed by TNBGD people in order to keep themselves safe online. The discussion then places these strategies in conversation with notions of digital citizenship and intersectional feminist digital citizenship.

Participants expressed concerns over the safety of the internet connection in their countries. A trans woman in Rwanda (R1) shared how for her, 'the internet is not good; it is not safe'. Another Rwandan trans woman (R8) shared her concerns with regards to government surveillance, sharing that 'you hear that the government can easily access your data'. While another trans woman from Rwanda shared that she felt that 'they are watching me, they are looking at my moves on the internet, it is not safe'. A Ugandan trans woman (U3) shared that 'sometimes people use your internet connection to track you ... They are tracking us through it ... When it comes to transgender people and the internet, nothing is safe'. While digital technologies make surveillance of TNBGD people possible, it is also through the very same digital technologies that TNBGD people may build counterpower in order to create spaces online that have different power relationships. The violence, such as surveillance and transphobia, that TNBGD people experience hinders their ability to fully participate in political citizenship. It is through the creation of digital safe spaces that are informed by a feminist ethics of care that TNBGD people are able to seek refuge, share their experiences, organize tactics of solidarity, and determine rights-claims that can be made as part of future feminist digital citizenship.

It is important to note that concerns regarding surveillance emerged from Rwanda and Uganda, two countries where TNBGD people are most at risk of government persecution. The fears shared here are not unfounded, given that social media platforms and apps have been used to track LGBTQIA+ people, such as in the aforementioned case of police in Egypt targeting a gay man through Grindr (Van Zyl and McLeanv 2021). The way in which digital technologies are used for surveillance means that vulnerable groups such as TNBGD people need to have 'secure basic protections that the general population might take for granted' (Ellis, 2022: 183). TNBGD people are 'often an afterthought when it comes to user privacy and regulations', for instance (Nigro, 2019). This then places TNBGD people at an added risk of violence, because the digital technologies they use to connect, build community and practice digital citizenship have not accounted for their vulnerability or how parties such as state governments may make use of TNBGD people's data (Kafer and Grinberg, 2019; Van Zyl and McLean, 2021).

The nature of the violence, as described by participants, could be categorized into four key forms: dismissal of gender identity, sharing of images without permission, hateful comments, and threats of violence and death. Deliberately disregarding someone else's identity based on their gender has been found to be linked to physical violence and also to have devastating effects on TNBGD individuals' social integration and mental health (Goldblum et al., 2021). The levels of self-harm recorded among TNBGD people are significantly higher than cisgender people due to the levels of violence and hate they are subjected to. Threats of violence and death to participants were particularly concerning.

TNBGD people are often the targets of threats of physical violence, which studies have shown have severe and harmful effects on their mental and physical health and their ability to live freely and safely (Goldblum et al., 2021). Types of threats often associated with OGBV are online threats of physical violence and assault with the intent to cause bodily harm, including sexual abuse. Trans women of colour are particularly vulnerable (Matsuzaka and Koch, 2019). Participants shared how they receive threats of violence, including threats of sexual and physical harm and death threats. The consequences of these threats of violence on the mental health of TNBGD people are severe. In a 2020 UK report on understanding the nature and the impact of transphobic hate crimes and prejudice, it was reported that '70% of respondents stated that transphobia had an impact on their mental health' and that 'nearly 50% of respondents said they had self-harmed and more than 50% had contemplated self-harm or suicide' (Bradley, 2020: 17). It is critical that we understand the severe consequences of OGBV on mental health and the risk it poses to the physical well-being of TNBGD people. It is also critical that we undertake similar impact studies for the Global South (see also Jauk, 2013). What is particularly concerning is not only the threat of violence but also how this violence impacts on TNBGD people's digital citizenship such that the fear of violence may result in self-censorship or non-participation in digital spaces.

One participant, a non-binary person from Botswana (B3), spoke of how they had been the target of OGBV after participating in a TNBGD trend online.[1] They explained how a few hours after posting their content online, their partner called them and asked them if they were okay before telling them that their content had gone viral and received national attention. This attention was not positive or affirming but rather transphobic and violent. It included 'pastors quoting the Bible, people quoting scientific theories [about gender]', as well as more severe threats. These treats included B3 being told that '"You need to be corrected" – people saying they are willing to violate me'. What B3 is referring to here is 'corrective rape', which is enacted by heterosexual men on lesbian women and gender-diverse people in order to address what they believe to be a divergence from gender roles (Koraan, 2015: 58). The participant described this as 'painful, honestly, to tell you the truth, because even my mom was freaking out. She didn't understand what non-binary meant, but she did understand that this is cyberbullying, and my child is not coping'.

A non-binary trans femme participant from South Africa (S1) had also received similar threats: 'I occasionally get death threats from anonymous email accounts. Which is wild, "There's something in my spam that doesn't look like spam. Oh, it's a death threat. Cool, cool, cool new email that I can block."' Another non-binary South African (S4) had also received comments on their posts about trans rights, sharing how 'you now start hearing, "transwomen are not women. I hate them. I would kill them."'

These threats of violence, as described above, are extremely distressing to the participants, and deeply concerning, especially given the graphic and gratuitous nature of the threats, including threats of sexual assault and murder. TNBGD people need better reporting mechanisms and responses from platforms to ensure their safety in order to ensure their continued participation in digital spaces.

Navigating, managing and remaining safe online

In order to continue participating and enacting feminist digital citizenship in the face of OGBV, TNBGD participants shared how they navigated and managed instances of OGBV and the strategies they employed to remain safe online in order to continue to gather information, connect with each other and actively participate in digital spaces. TNBGD individuals face a disproportionate amount of harassment, discrimination and violence online, and it is important to take steps to mitigate these risks and create a safer online environment (Noack-Lundberg et al., 2020; Haimson et al., 2020). Some participants spoke of using closed groups as a means to maintain community safety, while others spoke of using blocking – individually and collectively – as a safety strategy post-attack.

Use of closed groups

Participants spoke of making use of closed groups as a means of keeping the TNBGD community safe. Closed group is a function or option on platforms such as Facebook, as well as chat applications such as WhatsApp and Telegram. The use of this function is an active decision, informed by an ethics of care, to create a sense of safety online. This is a critical first step in creating safe space in order to build counterpower.

A trans woman from Rwanda (R3) shared how she feels that these groups 'are safe'. Participants attributed the reason for feeling safe to closed groups that only accept people through an invitation process by people known to them. The option of closed groups is a tactic used by TNBGD people to keep each other safe. A non-binary transmasculine person from Botswana (B2) explained how

> we found that if we have open groups, it allows for everybody to jump in and disrupt our home, what we're trying to create for ourselves … make sure that we monitor and put security settings around it … we don't want to be engaging people on those platforms because they will derail the kind of work that we're trying to do.

Here the use of the closed group functionality is a deliberate choice in which TNBGD people resist and mitigate the risk of OGBV and the function to create 'our home', where there is a sense of safety. This safe space becomes a space of support, advice, resources and community for its members (Horton, 2022).

In addition to closed groups, participants spoke of other forms of keeping themselves safe. These included the management of the information they reveal about themselves. A trans woman from Rwanda (R2) shared that 'I don't post anything when I know that it's going to make me suffer … I try to choose what I can post … I can hide some people [from viewing the post]'. A non-binary transmasculine South African (S8) managed their safety through privacy measures such as 'not posting my pronouns, my safety measures would just be not talking about it or not making any inclination that I am genderqueer. Unless I know it's safe'. Other participants shared how they made use of VPNs to give them a sense of anonymity and privacy.

These tactics, such as withholding pronouns or using VPNs, are an active navigation and utilization of digital technologies. They are instances of the strategic use of digital technologies in a way that works for the TNBGD person while protecting them from violence. This active engagement and navigation of digital technologies and social media platforms indicate feminist digital citizenship.

Blocking and reporting

TNBGD people's desire to manage digital spaces and access to their community and their personal information is largely due to poor responses from social media platforms to the violence TNBGD people face. Participants shared how when they blocked and reported accounts or posts, platforms responded in one of two ways to their reporting, either providing no feedback or outright dismissal of their experience. A trans woman from Botswana (B1) shared that 'as soon as I block the person, I never receive any other update regarding them or their presence'. Other participants shared how platforms outright dismissed their experiences or took action against them instead of the perpetrators. For instance, a non-binary trans femme South African (S1) shared how on '[X/]Twitter I've laid reports of what I thought was active harassment, but they thought it wasn't'. While another non-binary South African (S7) shared how in their case someone had commented on a post of theirs with threats, and the response from Facebook was to delete their post and 'I got my account restricted'.

Social media platforms perpetuate and are complicit in the violence TNBGD people experience. Platforms often fail to make visible their reporting mechanism. Not all participants were aware that there were blocking and reporting mechanisms on social media platforms. This was a concerning finding of the study given the pervasiveness of OGBV targeting TNBGD people. Blocking and reporting tools can help users flag inappropriate content, cyberbullying and other harmful online behaviours that are associated with OGBV (Gagoshashvili and Kaletsky, 2018; Edwards, 2022). Ensuring greater access and visibility of blocking and reporting mechanisms and tools would ensure that TNBGD people are better able to participate safely in digital spaces.

Participants who were aware of the reporting mechanisms on social media platforms spoke of blocking and reporting as a means of managing their safety online. While participants shared several accounts of content that they opted to block or report, what is of particular interest to this chapter is a blocking and reporting technique that is far more targeted and is referred to here as a solidarity strategy (Haimson et al., 2020; Pettas, 2022).

Solidarity strategies

A few participants spoke of an interesting strategy as a response to OGBV they were witnessing being experienced by other TNBGD people online. They spoke of mass blocking and reporting, which refers to the act of blocking and reporting multiple users or pieces of content on an online platform. This is done by marginalized people online for various reasons, such as to protect oneself from harassment and online violence, to report spam or abusive content, or to enforce community guidelines (Contreras, 2021). Mass blocking and reporting are considered to be acts of resistance to the relentless violence experienced by TNBGD people in digital spaces.

A non-binary participant from South Africa (S5) shared how when they see someone else being targeted, 'I report immediately, I don't question the degree of the transphobia. As soon as I feel like a trans person on the internet is in a vulnerable space and trapped in that engagement. I'll pull the plug on that immediately'. There is an unquestioning knowing here of the presence of violence that sees this participant respond from a space of solidarity. This sense of connectedness due to shared lived experiences where violence is so prevalent has this individual tuned into the potential risk another TNBGD person may experience. 'I'll pull the plug' indicates the know-how around mechanisms to report violence and counter the risk to the individual in question.

A trans man from Botswana (B5) explained how in his experience he no longer reads as transgender but rather 'passes' and makes use of this passing strategically to intervene when other TNBGD people are being targeted. Explaining that 'Yes, then people listen'. They listen because they assume he is a cisgender straight man who is calling them out on their transphobia. This is an interesting example of 'passing privilege' being used in order to protect fellow TNBGD people. This individual is aware of the privilege of being read as cisgender and uses it strategically in this digital context where his profile is read as that of a cisgender man in order to protect a fellow TNBGD person. This shows an awareness of digital assets, such as images, and how they are used in order to protect and intervene in instances of violence.

One of the most interesting examples is of a collective of more than six hundred people who respond as a group to reports of violence online. A non-binary person from South Africa (S3) shared her experience of belonging to a group that actively blocks individuals who harass and troll TNBGD people: 'several of us will be reporting at once ... for example, that group ... if I want to report something, I put it on that platform. That platform has over 600 people and we are all going to actively report that account, so it will go down'. The words 'actively report that account' suggest that this is a deliberate and strategic response, feminist digital citizenship in action in the face of violence.

They continued to share that 'I don't navigate them at an individual level. So, if there's someone who is harassing, trolling, hate speech, screenshots and put on the WhatsApp group, and then within a matter of like, 30 minutes, it's down. They've taken it down'. This is interesting commentary on the power of a digital collective in building counterpower in response to violence online. The use of screenshots as examples and the speed of the reporting shows an acute awareness of digital platforms and the power of the collective in advocating and organizing around what would have most probably gone ignored by the platform if the collective had not intervened. This is a strategic and coordinated response where one person is not burdened with trying to report an account alone and is a strong example of feminist digital citizenship enacting resistance and solidarity in the face of OGBV.

Conclusion and recommendations

TNBGD people are subjected to violence due to their gender identity and expression and turn to digital spaces such as social media platforms because they offer safety that offline spaces do not. However, while deemed safer, violence is still prevalent online. TNBGD people, however, have learned how to navigate and manage this violence in order to stay safe. Social media platforms largely ignore TNBGD people's reports of violence, and what has emerged in response to this is the solidarity strategy of collective organizing in order to mass report instances of violence. TNBGD people have learned the power of collective organizing through digital technologies and mechanisms to resist and mitigate against this violence. This organizing and resistance is an example of feminist digital citizenship in that TNBGD people actively make use of digital technologies and mechanisms to address power imbalances, mitigate risks against members of their community and respond from a space of collective solidarity to ensure that digital spaces remain safe for TNBGD people.

TNBGD people require safe digital spaces to gather, organize and establish a sense of community. Their experiences of OGBV must be taken seriously by key stakeholders, including social media platforms. This chapter makes recommendations for ensuring the safety of TNBGD people online. For TNBGD people, it is advised that they make use of the privacy settings available on their social media accounts to ensure their safety by controlling access to the viewing of posts, content and personal information. It is further recommended that TNBGD people are selective with the personal information that they opt to share, such as location, phone number or email address, to ensure that this information is not shared with unknown accounts or individuals. Should TNBGD people experience violence online, the blocking and reporting features on social media platforms can be used to prevent further contact from perpetrators.

Social media platforms can support TNBGD people by engaging with them to better understand their experiences and challenges in order to design more inclusive and safe platforms. Platforms should develop clear and comprehensive policies that prohibit online harassment, bullying and hate speech, which contribute towards OGBV, and ensure that these policies are enforced consistently and with care. In addition, platforms can provide inclusive and multilingual blocking and reporting tools for reporting violence, as well as creating educational materials about blocking, flagging and reporting, which may be in the style of a 'tour' of new features. Further, it is critical that social media platforms protect the privacy and personal information of TNBGD people, ensuring that their identities are not revealed without their consent, and commit to never selling their user data to any third party or sharing this data with governments or criminal justice systems.

Note

1 The details of the trend are not fully described here to protect the identity of the participant.

Bibliography

Ajaja, S. (2023), 'Feminist Digital Citizenship in Nigeria', in T. Roberts and T. Bosch (eds), *Digital Citizenship in Africa: Technologies of Agency and Repression*, London: Bloomsbury Academic, 117–47.

Andresen, M. (2020), 'Rwanda's Transgender Community Face Violent Detentions for Being Trans', *VICE*, 20 November, https://www.vice.com/en/article/4adxv9/rwandas-transgender-community-face-violent-detentions-for-being-trans.

Article 19 (2015), Right to Online Anonymity: Policy Brief, June, https://www.article19.org/data/files/medialibrary/38006/Anonymity_and_encryption_report_A5_final-web.pdf.

Association for Progressive Communications (2018), *Providing a Gender Lens in the Digital Age: APC Submission to the Office of the High Commissioner for Human Rights' Working Group on Business and Human Rights*. https://www.apc.org/en/node/35212.

Association for Progressive Communications (2017), *Online Gender-based Violence: A Submission from the Association for Progressive Communications to the United Nations Special Rapporteur on Violence Against Women, its Causes and Consequences*. https://www.apc.org/sites/default/files/APCSubmission_UNSR_VAW_GBV_0_0.pdf;.

Bradley, C. (2020), *Transphobic Hate Crime Report 2020*, Galop. https://galop.org.uk/resource/transphobic-hate-crime-report-2020.

Clarke, V., Braun, V. and Hayfield, N. (2006), 'Thematic Analysis', in J. A. Smith (ed.), *Qualitative Psychology: A Practical Guide to Research Methods*, London: SAGE Publications.

Contreras, B. (2021), '"I Need My Girlfriend Off TikTok": How Hackers Game Abuse-reporting Systems', *Los Angeles Times*, 3 December. https://www.latimes.com/business/technology/story/2021-12-03/inside-tiktoks-mass-reporting-problem.

Cooky, C., Linabary, J. R. and Corple, D. J. (2018), 'Navigating Big Data Dilemmas: Feminist Holistic Reflexivity in Social Media Research', *Big Data & Society*, 5 (2).

Crenshaw, K. (1991), 'Mapping the Margins: Intersectionality, Identity Politics, and Violence Against Women of Color', *Stanford Law Review*, 43, 1241–99.

Crenshaw, K. (1989), 'Demarginalizing the Intersection of Race and Sex: A Black Feminist Critique of Antidiscrimination Doctrine, Feminist Theory and Antiracist Politics', *University of Chicago Legal Forum*, 1, 139–67.

Edwards, C. (2022), 'Three Tools for Journalists to Protect Themselves Against Online Harassment', *Journalism.co.uk*, 2 August, https://www.journalism.co.uk/news/three-tools-for-journalists-to-protect-against-online-harassment/s2/a952257.

Edwards, R. and Mauthner, M. (2012), 'Ethics and Feminist Research: Theory and Practice', in T. Miller, M. Birch, M. Mauthner and J. Jessop (eds), *Ethics in Qualitative Research*, Second Edition. London: SAGE Publications.

Ellis, J. R. (2022), 'Blurred Consent and Redistributed Privacy: Owning LGBTQ Identity in Surveillance Capitalism', *Diversity in Criminology and Criminal Justice Studies*, 27, 183–96.

Epprecht, M. (2019), 'Botswana Recognizes LGBTQ Rights, Leading the Way in Southern Africa', *The Conversation*, 29 July, https://theconversation.com/botswana-recognizes-lgbtq-rights-leading-the-way-in-southern-africa-119277.

Faith, B. (2018), 'Tackling Online Gender-Based Violence', *Institute of Development Studies*, 3 December, https://www.ids.ac.uk/opinions/tackling-online-gender-based-violence.

Fawcett, B. and Hearn, J. (2004), 'Researching Others: Epistemology, Experience, Standpoints and Participation', *International Journal of Social Research Methodology*, 7 (3): 201–18.

Gagoshashvili, M. and Kaletsky, K. (2018), A Double-Edged Sword: The Internet as a Tool for Trans Activism', *Access Now*, 20 June, https://www.accessnow.org/a-double-edged-sword-the-internet-as-a-tool-for-trans-activism.

Gender DynamiX. (2021), *Annual Report*, https://www.genderdynamix.org.za/gdx-annual-reports.

Goldblum, P., Testa, R., Pflum, S., Hendricks, M., Bradford, J. and Bongar, B. (2021), The Relationship Between Gender-Based Victimization and Suicide Attempts in Transgender People, *Professional Psychology: Research and Practice*, 43 (5): 468–75.

Graaff, K. (2021), 'The Implications of a Narrow Understanding of Gender-Based Violence', *Feminist Encounters: A Journal of Critical Studies in Culture and Politics*, 5 (1): 1–12.

Guest, G., MacQueen, K. M. and Namey, E. E. (2012), *Applied Thematic Analysis*, London: SAGE Publications.

Gupta, A. (2019), 'Towards Participatory Democracy: Can Digitalisation Help Women in India?', *Indian Journal of Public Administration*, 65 (4): 897–915.

Haimson, O. L., Buss, J., Weinger, Z., Starks, D. L., Gorrell, D. and Baron, S. B. (2020), 'Trans Time: Safety, Privacy, and Content Warnings on a Transgender-Specific Social Media Site', *Proceedings of the ACM on Human-Computer Interaction*, 4 (CSCW2): 1–27.

Harding, S. (2004), 'Introduction: Standpoint Theory as a Site of Political, Philosophical, and Scientific Debate', in S. Harding (ed.), *The Feminist Standpoint Theory Reader: Intellectual and Political Controversies*. New York and London: Routledge.

Henry, N., Vasil, S. and Witt, A. (2022), 'Digital Citizenship in a Global Society: A Feminist Approach', *Feminist Media Studies*, 22 (8): 1972–89.

Henry, N. and Powell, A. (2018), 'Technology-Facilitated Sexual Violence: A Literature Review of Empirical Research', *Trauma, Violence, & Abuse*, 19 (2): 195–208.

Hesse-Biber, S. N. (ed.) (2007), *Handbook of Feminist Research: Theory and Praxis*, London: SAGE Publications.

hooks, b. (2000), *Feminism is for Everybody: Passionate Politics*, London: Pluto Press.

Horton, C. (2022), 'Gender Minority Stress in Education: Protecting Trans Children's Mental Health in UK Schools', *International Journal of Transgender Health*, 24 (2): 195–211.

Human Rights Campaign (2023), The Epidemic of Violence Against the Transgender and Gender Non-Conforming Community in the United States: The 2023 Report. https://reports.hrc.org/an-epidemic-of-violence-2023.

Human Rights Campaign (2020), Fatal Violence Against the Transgender and Gender Non-Conforming Community in 2020. https://www.hrc.org/resources/violence-against-the-trans-and-gender-non-conforming-community-in-2020.

Hundle, A. K., Szeman, I. and Hoare, J. P. (2019), 'What is the Transnational in Transnational Feminist Research?', *Feminist Review*, 121 (1): 3–8.

Igonya, E. K. (2022), 'Rwanda: LGBT Rights are Protected on Paper but Discrimination and Homophobia Persist', *The Conversation*, 16 May. https://theconversation.com/rwanda-lgbt-rights-are-protected-on-paper-but-discrimination-and-homophobia-persist-182949.

Iranti (2019), *Data Collection and Reporting on Violence Perpetrated Against LGBTQI Persons in Botswana, Kenya, Malawi, South Africa and Uganda: A Report Prepared for* the Arcus Foundation. https://www.arcusfoundation.org/wp-content/uploads/2020/04/Iranti-Violence-Against-LGBTQI-Persons-in-Botswana-Kenya-Malawi-South-Africa-Uganda.pdf.

Jauk, D. (2013), Gender Violence Revisited: Lessons from Violent Victimization of Transgender Identified Individuals', *Sexualities*, 16 (7): 807–25.

Johnstonbaugh, M. (2021), 'Men Find Trophies Where Women Find Insults: Sharing Nude Images of Others as Collective Rituals of Sexual Pursuit and Rejection', *Gender & Society*, 35 (5): 665–90.

Jordan, S. P., Mehrotra, G. R. and Fujikawa, K. A. (2020), 'Mandating Inclusion: Critical Trans Perspectives on Domestic and Sexual Violence Advocacy', *Violence Against Women*, 26 (6-7): 531–54.

Kafer, G. and Grinberg, D. (2019), 'Queer Surveillance', *Surveillance & Society*, 17 (5): 592–601.

Knapp, D. v. B. and Ogunbanjo, G. (2009), 'Feminism and the Ethics of Care', *South African Family Practice*, 51 (2): 116–18.

Lanham, M., Ridgeway, K., Dayton, R., Castillo, B. M., Brennan, C., Davis, D. A., Emmanuel, D., Morales, G. J., Cheririser, C., Rodriguez, B., Cooke, J., Santi, K. and Evens, E. (2019), '"We're Going to Leave You for Last, Because of How You Are": Transgender Women's Experiences of Gender-Based Violence in Healthcare, Education, and Police Encounters in Latin America and the Caribbean', *Violence and Gender*, 6 (1): 37–46.

Laskey, P., Bates, E. A. and Taylor, J. C. (2019), 'A Systematic Literature Review of Intimate Partner Violence Victimisation: An Inclusive Review Across Gender and Sexuality', *Aggression and Violent Behavior*, 47, 1–11.

Lee, P. W. Y., Lynch, I. and Clayton, M. (2013), 'Your Hate Won't Change Us! Resisting Homophobic and Transphobic Violence as Forms of Patriarchal Social Control', *Triangle Project Human Rights Report*.

Lombardi, E. L., Wilchins, R. A., Priesing, D. and Malouf, D., (2002), 'Gender Violence: Transgender Experiences with Violence and Discrimination', *Journal of Homosexuality*, 42 (1): 89–101.

Matsuno, E., Bricker, N. L., Savarese, E., Mohr Jr. R., and Balsam, K. F. (2022), 'The Default is Just Going to be Getting Misgendered: Minority Stress Experiences Among Nonbinary Adults', *Psychology of Sexual Orientation and Gender Diversity*, 11 (2): 202–14.

Matsuzaka, S. and Koch, D. E. (2019), 'Trans Feminine Sexual Violence Experiences: The Intersection of Transphobia and Misogyny', *Affilia*, 34 (1): 28–47.

McLean, N. and and Cicero, T. (2023), 'The Left Out Project: The Case for an Online Gender-Based Violence Framework Inclusive of Transgender, Non-Binary and Gender-Diverse Experiences', *The Feminist Internet Research Network, Association for Progressive Communications*. https://firn.genderit.org/research/left-out-project-report-case-online-gender-based-violence-framework-inclusive-transgender.

McLean, N. (2022), *Feminist Internet Research Network: Meta-research project report*. Association for Progressive Communications. https://www.apc.org/en/node/38022.

McLean, N. (2020), 'Creating Safe Spaces: Digital as an Enabling Environment for TNB People', in S. K. Kattari (ed.), *Social Work and Health Care Practice with Transgender and Nonbinary Individuals and Communities*, New York and London: Routledge, 331–42.

McLean, N. (2018), *The Rupture in the Rainbow: An Exploration of Joburg Pride's Fragmentation, 1990 to 2013*, Doctoral thesis, Rhodes University.

McLean, N. (2015), Zwischen Party und Protest, *Sudlink*, 174, 17.
McLean, N. and Mugo, T. K. (2015), 'The Digital Age: A Feminist Future for the Queer African Woman', *IDS Bulletin*, 46 (4): 97–100.
Mhaka, T. (2021), 'Africa's LGBTQ Communities Need More Protection and Support', *Al Jazeera*, 30 April. https://www.aljazeera.com/opinions/2021/4/30/africas-lgbtq-communities-need-more-protection-and-support.
Mudavanhu, S. and Radloff, J. (2013), 'Taking Feminist Activism Online: Reflections on the "Keep Saartjie Baartman Centre Open" e-campaign', *Gender & Development*, 21 (2): 327–41.
Müller, A., Daskilewicz, K., Kabwe, M. L., Mmolai-Chalmers, A., Morroni, C., Muparamoto, N., Muula, A. S., Odira, V., Zimba, M. and Southern and Eastern African Research Collective for Health (SEARCH) (2021), 'Experience of and Factors Associated With Violence Against Sexual and Gender Minorities in Nine African Countries: A Cross-Sectional Study', *BMC Public Health*, 21, 1–11.
Nigro, A. (2019), 'Why Shaky Data Security Protocols for Apps Put LGBTQ People at Risk', *Pacific Standard*, 21 May. https://psmag.com/social-justice/why-data-security-is-essential-for-the-lgbtq-community.
Noack-Lundberg, K., Liamputtong, P., Marjadi, B., Ussher, J., Perz, J., Schmied, V., Dune, T. and Brook, E. (2020), 'Sexual Violence and Safety: The Narratives of Transwomen in Online Forums', *Culture, Health & Sexuality*, 22 (6,) 646–59.
Noble, S.U. (2018), 'Algorithms of Oppression: How Search Engines Reinforce Racism', in *Algorithms of Oppression*, New York: New York University Press.
Nyeko, O. (2022), 'Uganda Bans Prominent LGBTQ Rights Group', *Human Rights Watch*, 12 August. https://www.hrw.org/news/2022/08/12/uganda-bans-prominent-lgbtq-rights-group.
Pereira, C. (2017), 'Feminists Organising—Strategy, Voice, Power', *Feminist Africa*, 22, 16–30.
Pettas, D., Arampatzi, A. and Dagkouli-Kyriakoglou, M. (2022), 'LGBTQ+ Housing Vulnerability in Greece: Intersectionality, Coping Strategies and the Role of Solidarity Networks', *Housing Studies*, 26 June, 1–19.
Ponnam, A. and Dawra, J. (2013), 'Discerning Product Benefits Through Visual Thematic Analysis', *Journal of Product & Brand Management*, 22 (1): 30–9.
Posetti, J., Shabbir, N., Douglas, O. and Gardiner, B., (2022), 'The Chilling: A Global Study of Online Violence Against Women Journalists', *UNESCO: United Nations Educational, Scientific and Cultural Organisation. France*. Retrieved from https://coilink.org/20.500.12592/wgvd03 on 21 Aug 2024. COI: 20.500.12592/wgvd03.
Powell, A., Scott, A. J. and Henry, N. (2020), 'Digital Harassment and Abuse: Experiences of Sexuality and Gender Minority Adults', *European Journal of Criminology*, 17 (2): 199–223.
Preissle, J. (2007), 'Feminist Research Ethics', in S. N. Hesse-Biber (ed.), *Handbook of Feminist Research: Theory and Praxis*. London: SAGE Publications.
Reid, G. (2020), 'LGBTQ Inequality and Vulnerability in the Pandemic', in *Human Rights Watch*, Retrieved 29 March 2021 from https://tinyurl.com/reidg2020
Reid, G. (2017), 'Victory for Gender Identity in Botswana: Country's High Court Rules in Favor of Transgender Man', *Human Rights Watch*, 3 October. https://www.hrw.org/news/2017/10/03/victory-gender-identity-botswana.
Roberts, T. and Bosch, T. (2023), *Digital Citizenship in Africa: Technologies of Agency and Repression*, London: Bloomsbury Academic.

Rodríguez-Dorans, E. (2018), 'Reflexivity and Ethical Research Practice While Interviewing on Sexual Topics', *International Journal of Social Research Methodology*, 21 (6): 747–60

Rogers, M. M. (2021), 'Exploring the Domestic Abuse Narratives of Trans and Nonbinary People and the Role of Cisgenderism in Identity Abuse, Misgendering, and Pathologizing', *Violence Against Women*, 27 (12-13): 2187–207.

SAnews (2021), 'Spate of Attacks on LGBTQI+ Community in SA', 1 July. https://www.sanews.gov.za/south-africa/spate-attacks-lgbtqi-community-sa.

Schudson, Z. and van Anders, S. (2019), '"You Have to Coin New Things": Sexual and Gender Identity Discourses in Asexual, Queer, and/or Trans Young People's Networked Counterpublics', *Psychology & Sexuality*, 10 (4): 354–68.

Staples, J. M., Bird, E. R., Masters, T. N. and George, W. H. (2018), 'Considerations for Culturally Sensitive Research with Transgender Adults: A Qualitative Analysis', *The Journal of Sex Research*, 55 (8): 1065–7.

Suzor, N., Dragiewicz, M., Harris, B., Gillet, R., Burgess, J. and Van Geelen, T. (2018), 'Human Rights by Design: The Responsibilities of Social Media Platforms to Address Gender-Based Violence Online', *Policy & Internet*, 11 (1): 84–103.

Teal, J. and Conover-Williams, M. (2016), 'Homophobia without Homophobes: Deconstructing the Public Discourses of 21st Century Queer Sexualities in the United States', *Humboldt Journal of Social Relations*, 38, 12–27.

The Constitution of the Republic of South Africa, Act 108 of 1996.

Thoreson, R. (2008), 'Somewhere Over the Rainbow Nation: Gay, Lesbian and Bisexual Activism in South Africa', *Journal of Southern African Studies*, 34 (3): 679–97.

United Kingdom Government (2022), 'Country Policy and Information Note: Sexual Orientation and Gender, Uganda', 11 February. https://www.gov.uk/government/publications/uganda-country-policy-and-information-notes/country-policy-and-information-note-sexual-orientation-and-gender-uganda-february-2022-accessible-version.

USAID (2019), *Rwanda: Gender and Social Inclusion Analysis Report*, United States Agency for International Development.

Van Zyl, I. and McLean, N. (2021), 'The Ethical Implications of Digital Contact Tracing for LGBTQIA+ Communities', *Proceedings of the 1st Virtual Conference on Implications of Information and Digital Technologies for Development.*

Web Foundation (2020), 'The Impact of Online Gender-Based Violence on Women in Public Life', 25 November. https://webfoundation.org/2020/11/the-impact-of-online-gender-based-violence-on-women-in-public-life.

Wells, H. and Polders, L. (2006), 'Anti-Gay Hate Crimes in South Africa: Prevalence, Reporting Practices, and Experiences of the Police', *African Feminisms*, 2 (3): 20–8.

Wigginton, B. and Lafrance, M. N. (2019), Learning Critical Feminist Research: A Brief Introduction to Feminist Epistemologies and Methodologies', *Feminism & Psychology*, 0 (0): 1–17.

Yun, J. Y. (2020), 'Feminist Net-Activism as a New Type of Actor-Network that Creates Feminist Citizenship', *Asian Women*, 36 (4): 45–65.

Chapter 5

DIGITAL FRUGALITY IN LOW-INCOME COMMUNITIES IN SOUTH AFRICA: ENABLING WOMEN'S CITIZENSHIP OF SURVIVAL

Alette Schoon and Marion Walton

Survival, feminism and the mobile diary method

The most notable act of feminist citizenship in South African history is the Women's March of 1956, when thousands of Black women from all over the country came together at the seat of government of the apartheid regime to protest the racist pass laws. They held posters and sang and shouted slogans, but the most notable slogan they coined was 'wathinta umfazi wathinta imbokodo!' roughly translated as 'you strike a woman, you strike a rock'. The rock referred to the grinding stone, a cooking implement of traditional African women. That slogan celebrates what might otherwise be conceptualized as a passive victimhood and instead claims for women a special power: the ability to survive injustice, to endure hardship and to create lives that are not defined by suffering. Drawing on notions of African grandmothers', mothers' and daughters' survivalist knowledge (Motsemme, 2011; Magoqwane, 2018) this chapter argues for an African feminist digital citizenship that acknowledges women's struggles and their survival strategies. The historic Women's March embraced direct political action and foregrounded resilience and survival as skills that demand collective agency and survivalist knowledge.

Women's resilience is still recognized and celebrated through digital media in the twenty-first century in South Africa. During the time of our study, the #FeesMustFall protests had a massive national impact, lowering the costs of university education. During these protests, social media became a tool for radical activism (Bosch, 2017). Student movement social media campaigns highlighted issues of rape and patriarchy on campuses, challenged the gendered violence at work in protests and produced a cohort of powerful women student leaders (Dlamini, 2022; Maluleke and Moyer, 2020). These student leaders embraced the slogan of the 1956 protests, calling themselves 'Mbokodo' and using the #MbokodoLeads hashtag as a reference to the ability of African women to endure any hardship without it leaving a mark (Dlamini, 2022).

Due to the ubiquitous nature of mobile phones, these devices have become integral to everyday life and everyday struggles for survival. For this study, mobile phone practices were investigated at two research sites of precarious survival: an impoverished ward of a Cape Town township and various rural villages near a small coastal town in the former apartheid 'homeland' of Ciskei. All the participants were navigating frugal economic conditions of poverty, which matched global poverty indexes. These were both people classified as within the upper bound poverty line where their household income was deemed just able to cover a range of general minimum living costs, such as transport, housing, electricity and food (which adjusted to 2016 prices is R 1,031 or USD 76 per month per person), as well as those officially below the food poverty line (then R 335 or USD 25 per month per person) who were barely able to subsist, as their household income only covered the minimum food expenses deemed necessary for survival (STATSSA, 2015). All participants were Black South Africans. Due to the cumulative dispossession of colonialism and apartheid, Black South Africans still disproportionally bear the brunt of poverty. While the advent of democracy has had a significant impact on the growth of the Black middle class, South African society remains one of the most unequal in the world and, for most Black South Africans, poor Black neighbourhoods with few opportunities remain the reality (Seekings, 2008, 2015).The challenges our participants faced are thus not minority concerns but shared by the majority of South Africans. Poverty statistics reveal that 51 per cent of South Africans fall below global poverty metrics, relying on an income that sustains the bare basics of life, including food, housing, clothing and transport, while 20 per cent fall below the lower poverty metrics that only sustain basic food intake.

Yet several studies have revealed that even in such constrained environments, and despite South Africa's relatively high data and voice connectivity costs, many people in this low income bracket still find ways to be connected intermittently, not just to mobile phone providers but also to the mobile internet (Donner and Gitau, 2009; Schoon, 2014; Schoon and Strelitz, 2014; Walton, Haßreiter and Marsden, 2012; Walton and Donner, 2012). While such degrees of connectivity and struggles around access have indeed been encountered in pockets of marginalized people in the Global North (Faith, 2018), such limited connectivity is arguably quite exceptional in such contexts. In contrast, in the Global South, intermittent connectivity is the norm, not the exception, and the 'always on' internet is constrained to minority elite 'middle class'[1] spaces (Donner, 2015). This makes these 'less-connected' contexts especially significant for understanding digital citizenship among ordinary Africans. Less-connected users tend to use the mobile internet intermittently, primarily on text-based messaging apps, if they use the mobile internet at all, since many still rely primarily on mobile voice telephony. Researching these practices requires different digital methods, as such types of connectivity might not leave readily available traces for researchers to study online. The study of digital practices on the African continent differs so radically from what is considered the norm in the developed world that it requires careful qualitative research to render people's agency visible (Schoon et al., 2020). Even in seemingly 'disconnected' contexts, many people manage and engage with (dis)connectivity in complex ways (De Lanerolle, Schoon and Walton, 2020).

For this reason, we developed a mobile diary method (De Lanerolle, Schoon and Walton, 2020). Through in-depth interviews conducted by research assistants in their home language and joint overviews of mobile phone records, our participants revealed the minute details of their lives during the previous day, allowing researchers to construct a detailed contextual record of mobile practices. We were thus able to root the study in the rich empirical details of a recent past, exploring how mobile phone use was embedded in the practices of everyday life, thus following a media practice approach (Couldry, 2004), and through foregrounding the most marginalized voices, enabling a radical decentring (Schoon et al., 2020). Our study was not particularly focused on women but engaged both men and women to reflect on yesterday ('izolo' in isiXhosa) in such mobile diaries. Participants shared information about their devices, their network providers and their use of airtime, data and apps, often consulting the digital logs left by the previous day's mobile activities to help them remember details such as messages and phone calls made and received. Interviewers sought to understand what these practices meant to the participants, thus 'thickening' the digital trace data recorded (and often erased from) the phones (Latzko-Toth, Millette and Bonneau, 2016).

Despite our mobile diary research project not being focused on women, the role women played as connectors, carers and innovators of both digital systems and communities emerged so powerfully from our data that it prompted us to write this chapter. Our findings need to be considered in the context of the particular challenges faced by women in general in South Africa as well as the intersectional challenges of being a poor Black woman. Violence against women is pervasive, and South Africa has one of the highest rape statistics in the world (Gqola, 2007). Black women are the least represented in the formal labour sector. Low-income women whose families depend on social welfare are put into a particularly challenging position, as in South Africa, the social welfare system is designed to consider the family the locus of care and relies on social grants (Sevenhuijsen et al., 2003). These grants are disseminated to women as the presumed primary caregivers in the home, who are assumed to be able to rely on a male partner to supplement their income; therefore, the grants are quite modest. Many households in low-income neighbourhoods are headed by women, but since social grants are not large enough to match even the lower poverty metrics, women have to engage in the labour market as well as doing primary care work, placing an incredibly difficult burden on them (Sevenhuijsen et al., 2003). Gouws (2016) argues that this contradictory way that the state has positioned women in South Africa as equal to men in the transnational context, but then in the local context as central to care, has created citizenship paradoxes. Women may have equal rights in some spaces, but as the official providers of care they have many more responsibilities.

The minute details of women's lives observed in these mobile diaries rendered visible the material realities of women's domestic labour and care, which might not have emerged otherwise. However, our mobile diary method focused on mobile communication and did not foreground feminist concerns, such as exploring

GBV in the home. We may have therefore certainly missed many important details of participants' lives that our study was not initially set up to explore, or which participants hesitated to share. As two white women in South Africa, we were acutely aware that the world revealed by these diaries was indeed very far from our own daily realities of racialized privilege, relative security and access to education, employment and home ownership. We therefore recruited multilingual research assistants who were fluent in isiXhosa and had the linguistic and cultural knowledge to conduct nuanced interviews and produce translations. Our insights are indebted to these research assistants, the women participants who made time to answer our questions and the conceptual work of African feminists, such as Babalwa Magoqwana (2018) and Nthabiseng Motsemme (2011), which informed our analysis and emerged out of intimate knowledge and lived experience of growing up in such communities.

The stories that emerged through the mobile diaries reveal the agency of South African women in making both their communities and their digital technologies work. We set out some of these stories below through several case studies. In doing so, we hope to argue that the mobile practices we observed need to be considered as integral to a digital citizenship rooted not only in care, but in survival. We were particularly struck by the parallels between the agency needed to constitute the conditions for the survival of families and the agency needed to maintain digital connectivity. In this chapter, we ask: how do ordinary South African women's labour and communal knowledge enable survival and digital connectivity? In spaces where both institutions and digital public spaces for deliberation are largely absent, how might we conceptually link this to digital feminist citizenship?

Case studies from women's mobile diaries

We selected five case studies from our mobile diary interviews to explore ways in which women's everyday digital practices build and maintain the connectedness required for survival. This includes learning the skills of digital frugality, doing the safety work needed to carve out spaces for privacy, and expressing what we argue can be considered a feminist ethos of mutual support.

Zintle's uses of digital frugality to sustain her mobile connectivity

In the context of scarcity and precarity, mobile connectivity requires a set of practices and the knowledge to create and preserve digital infrastructures. These practices and knowledge, which we call 'digital frugality', are not acknowledged in the literature around digital divides, which tends to consider technology and infrastructure as static and fixed resources that should be accessed and mastered in the ways their designers envisaged. Instead we posit digital frugality as an active crafting of 'lively infrastructure' (Amin, 2014) based in the agency and ingenuity of the poor and the marginalized of making do, cobbling together and sharing techniques emerging from knowledge developed in constant experimentation, often fuelled by desperation.

Zintle was an unemployed young woman who stayed at home with her toddler, mother and aunts in a rural Eastern Cape Village. Infrastructure in the village had been neglected under apartheid's racially segregated provision, and mobile connectivity was still tenuous and constantly threatened. On the day we spoke to Zintle, a young man she knew had been caught stealing electricity poles on her street. Such vandalism and theft intensified the structural inequalities that beset the village, making connectivity challenging. Nonetheless, Zintle depended on mobile connectivity to access municipal opportunities, share advice with women's groups and maintain her relationship of support with her 'baby daddy'. She actively crafted such connectivity through her digital frugality. For example, on the day of the interview, she received a call from the local municipal councillor, who facilitated opportunities for low-paying internships through the government's public works programme, a strategy to address youth unemployment (Leech, 2024). Zintle never left her phone unattended, as she was always hoping the councillor might call with news of an opportunity.

She also needed to always be contactable by her baby daddy, who supported them financially. She did not want him to phone and to find her number engaged, as this might create questions around her loyalty to him. Consequently, she employed a range of strategies to ensure uninterrupted connectivity. She had two phones, and the baby daddy was the only one who had the number for her smartphone. Even though she used WhatsApp on the smartphone, she had set it up so that it used her other phone's number, keeping the smartphone number private. Besides sending WhatsApps to her baby daddy, Zintle would occasionally visit Facebook.

Mobile data was expensive, but she valued the advice she received from a Facebook group for women, which appeared as focused on fashion as it was on safety. She read out some of the advice to us: 'Don't go there without any money, maybe money to come back home, don't only take money to go there, so that if you find trouble there you'll have money to go home'. The smartphone was a hand-me-down from her brother, as it 'eats airtime' since there were no apparent controls for restricting data usage. For this reason, Zintle was mindful about switching on her mobile data settings only when she needed it, or as she put it 'I am not a frequent flyer on the internet'. To enable her intermittent connectivity, Zintle, like many of the other participants in our study, had an encyclopaedic knowledge of service provider specials and used two different service providers. Since she never made calls on the smartphone herself, she had a contract with Cell C, which had much more affordable data bundles, while the other phone used MTN, which had good specials on airtime.

Digital citizenship here therefore depends on practising digital frugality, as Zintle's digital activities of citizenship clearly illustrate: both her participation in the municipal employment program and the Facebook women's advice group depended on her actively controlling her connectivity, practising digital frugality.

Many other women shared their practices of digital frugality with us, describing ways to stretch connectivity through their knowledge of mobile service provider specials and sharing devices or online accounts or even combining devices to extend the usage of old phone chargers. It's important to understand that the way we define digital frugality here is not simply the arbitrary cutting back of data usage. It is a selective and strategic knowledge-based practice of stretching connectivity by deciding what digital activities to prioritize, where such decisions are rooted in women's knowledge of survival. The importance of such digital frugality will become evident from our other case studies, which show how intermittent connectivity is integral to a citizenship of survival.

Ma Fezeka sustains the social fabric with communicative work

Ma Fezeka's story demonstrates the importance of women's role as communicators in maintaining the fabric of a community and, indeed, in enabling civic actions such as protests. Ma Fezeka was an older woman who shared her home with her adult children and grandchildren in one of Cape Town's more established townships. Ma Fezeka did not shy away from expressing her opinions and was acutely aware of the reasons for the housing protests that had happened the day before we met with her. She was convinced that none of the residents of her ward had received home ownership because of government corruption. Although her three adult children qualified for housing and had been on the housing waiting list for a long time, they had again been passed over in the latest round of allocations. In her focus on the injustices encountered by her children, we can see how these strong feelings about rights and citizenship are informed by democratic discourses about the right to a home and her identity as a mother.

> We need more houses. My kids live in the backroom shacks. There are three of them living in that shack because they can't get houses ... it's their right to get decent housing.

Historically, South African women's civic activism was primarily informed by motherism (Ruddick, 1980), uniting women through a shared concern for their children (Fester, 2005). Motherism is a type of political mobilization where women primarily identify as mothers, which Fester (2005) argues has produced many successful political campaigns in the Global South.

Despite empathy for protesters and concern about her children, Ma Fezeka had not joined the protest action. As elsewhere in South Africa, people had lost faith in deliberation, and protests had increasingly become chaotic and violent (Robins, Cornwall and von Lieres, 2008). At dawn, backyarders had blocked the access roads to the township, bringing protesters, police and local gangs into conflict. Ma Fezeka's daughters had woken her early, seeking refuge in her home, unable to go to work. The community's post office was destroyed, and small stores owned by Somali migrants were looted. If the initial aim of the protest was a collective claim on both housing and citizenship, these destructive gestures produced its reverse, revealing a 'fault line' (Ojebode et al., 2023) along ethnic lines between protesters and, in this case, immigrant Somali shop owners. This fracture of nationality added to longstanding tensions between 'backyarders', the families of long-term residents like Ma Fezeka, and more recent arrivals in the area living in informal settlements.

Ethno-religious fault lines have frequently spurred conflict in Africa (Ojebode et al., 2023). Roberts and Bosch (2023) argue that one should incorporate such notions of ethnic belonging into a definition of African citizenship that is primarily defined through the concepts of agency, rights and resistance. Such a public rights-based notion of citizenship, even if incorporating notions of ethnicity, still renders the important roles many women play in society invisible. Care work, such as raising children and looking after the sick and the disabled, should be recognized as a civic service (Pedersen, 1990). Citizenship's origins in notions of public civic action have always undermined women, as it has privileged the notion of citizen-as-earner over citizen-as-carer (Hernes, 1987). Lister (2003) argues that a feminist notion of citizenship should place notions of care, not rights, at its centre.

In this chapter, we similarly call for a recognition of care as being at the centre of African citizenship. This definition is arguably more important in the African context considering the greater importance of citizen's care work in societies where difficult socio-economic conditions and overstretched public institutions are the norm. Ma Fezeka's role in the protest can be conceptualized as based in such a citizenship of care. She empathized with the shop owners. She was disappointed by the destruction, and her mobile phone use centred on filling in for missing workers by cooking at a local crèche, passing messages between family members and taking care of a sick child. Everyday urgent communicative and caring responsibilities were intensified by the disruptive protest and violent police response.

The terrifying events of Ma Fezeka's day illustrate how protests do not guarantee safe participation for women. Feminist digital citizenship on the continent champions online spaces as a safer alternative space for digitally connected women to raise their voices (Ajaja, 2023).

In South Africa, digital feminist campaigns to highlight violence show women claiming such online protest spaces, resulting in hashtags such as #RUReferenceList, #RememberKhwezi and #AmINext (Maluleke, 2024; Maluleke and Moyer, 2020). However, such modes of feminist digital citizenship are still not accessible to everyone. For participants in our study, online space was not an alternative civic space, as they mostly did not have mobile data to participate. Political talk happened in spaces like the park, where some men from Ma Fezeka's neighbourhood told us they had discussed eyewitness videos of the protests while playing dominoes. No social media platforms were used; they simply passed around someone's actual mobile phone to watch the videos, creating a 'pavement internet' where digital media is shared offline and commented on face to face (Walton, 2014), leaving no digital traces and thus rendering such digital citizenship invisible to outsiders. The men were indeed unaware of the social media hashtags used by middle-class Capetonians to comment on the protests. Ma Fezeka did not even have access to such pavement internet, as she did not have the leisure to play dominoes and watch videos. Owing to the time-consuming labours of care, women often do not have time for overt civic involvement, thus rendering the politics of time an important element of feminist citizenship (Lister, 2003). Ma Fezeka's absence from public protest actions and discussions might render her actions invisible in a rights-based notion of citizenship. Yet her acts of care helped the crèche and family members cope with the protest, mending fault lines deepened by the violence. Women like her arguably ensured that the protest could continue and be less disruptive than it might have been, and their actions also helped restore relationships in the area.

Ma Fezeka's communicative efforts highlight the role of women's talk in weaving a caring citizenship into the social fabric. Feminists argue that women's talk accomplishes the work of maintaining relationships, performing activities in the community and enacting processes of care (Rakow, 1992: 10). These communicative practices are gendered work, a wide range of socially oriented communication, which is naturalized and often socially expected from women. These everyday forms of talk, (including denigrated stereotypically feminine genres such as so-called gossip) are crucial to the renewal and maintenance of networks and communities (Fischer, 1992: 212; Rakow, 1992: 10). Much like housework and childcare, they are invisible forms of labour that need to be made visible and their value recognized (Marwick, 2023: 106). Liberal citizenships focused on the expression of rights in public spaces would not recognize Ma Fezeka's communicative labour to mitigate the chaos created by the housing protest and manage her extended family's contingency plans. Feminist perspectives focused solely on digital actions on public platforms are also likely to miss the high stakes of caring work in marginal contexts like Ma Fezeka's, where public goods are unreliable and tenuous.

Ma Xoliswa creates a responsive community of care

Our second case study concerns another grandmother, living in a small village in a former homeland in the rural Eastern Cape. This case study highlights digital practices that might appear passive and lacking in agency, such as ensuring that one is always contactable, as is indeed central to being a responsive caregiver. The ethics of care recognizes that being responsive and attentive to the needs of others is a crucial skill for enabling care (Held, 2006). Ma Xoliswa's responsiveness therefore presents a different approach to communicative labour.

Ma Xoliswa was the sole breadwinner for two unemployed adult daughters who lived in the outside shack with their two toddlers and another three pre-teen grandchildren who lived with her in the main house. She kept chickens and pigs and planted vegetables on a communal plot, but mostly she earned her income from sewing. At the whirring sewing machine that dominated her modest living space, she would constantly reach into the pocket of her apron to feel for her tiny basic mobile phone to check if it was still there. In setting out her mobile diary she was adamant that she didn't make calls, she only received them, but she was always contactable. She ensured that her phone was always charged so that she was ready if there was any emergency in the family, such as a road accident. She did not understand men in the community who seemed not to care about their families and who would easily lose their phones when they were drunk.

Whereas Ma Fezeka was actively reaching out to everyone in the city, Ma Xoliswa might appear passive, simply waiting for others to call. If agency was a central feature of citizenship (Roberts and Bosch, 2023), did simply being contactable count? That day someone had called Ma Xoliswa to borrow her wheelbarrow to go to the shop, and the teacher who was absent from school had sent an SMS with homework for one of the grandchildren. Ma Xoliswa's fielding of calls might seem trivial, but she also took more urgent calls when her adult children were arrested for transgressing fishing quotas.

> Ma Xoliswa: Maybe my children are walking at the beach and they call me and say, 'Mama we encountered inspectors and we were detained,' you understand? It's helpful in that way, when you land into trouble.

She explained that men were unlikely to attend to such calls and would probably say 'No, just leave them'. Instead, she insisted, she would be actively involved if this happened. While she appeared to just be an ordinary grandmother, she revealed that she actually had substantial authority, as she had become the village contact for government fishing permits. She came to adopt this position due to her expertise in fishing and was one of the first people in the village to have a fishing permit. Indeed, the only call Ma Xoliswa admitted to making in her mobile diary was a call to her sister whose workplace has a view of the beach to ask what the sea was like to decide whether she should go fishing.

Ma Xoliswa's knowledge of fishing was a knowledge born from survival. When her children were small, her husband left her. She did not allow her children 'to have hungry tummy' and worked as a labourer in her neighbours' fields 'even threshing the corn to use the husks for pig feed'. During that time, fishing regulations were not enforced, and as a young mother, Ma Xoliswa would often wander into the rock pools and go hunting for the octopus, 'ulwandle', plucking one from underwater. Her embodied knowledge of fishing was evident from the way she mimed how the creature would stick to her arm with its suckers until she had plucked its intestines from its head, after which it would 'drop like a snake'. She exchanged the octopus meat for sugar and vegetables for her children.

Ma Xoliswa's story illustrates how the private concerns that emerge from caring for loved ones are often what propels women into active involvement in civic spaces (Lister, 2003). Ma Xoliswa's awareness of the importance of mothers like herself being able to fish to feed their children resulted in her championing subsistence fishing rights and holding a leadership position on this issue. In this way, citizenship is not independent of, but rooted in the challenges of everyday private life in the home and the need to care to collectively survive. It is from the home that struggles around rights emerge, but also where people cultivate dispositions for approaching challenges.

It is important to recognize that survival in the context of poverty does not only require overcoming material deprivation, but especially in deeply stratified societies, it also requires overcoming the psychological difficulties of shame and rage produced by inequality and maintaining a sense of dignity (Richardson and Skott-Myhre, 2012). Magoqwana (2018) argues that in South Africa, grandmothers are key to providing the support for communities to thrive; therefore, they should be recognized as an institution. UMakhulu (grandmothers, or literally 'the great ones') collectively provide the survivalist knowledge and structure to make life bearable and produce new generations of loved and hopeful young people in South Africa. Such ideas are rooted in Motsemme's (2011) call to resist representing African women as passive and to recognize the agency and knowledge embedded in their survivalist practices.

The prayer group builds a space of psychological care

Our next case study, of the rural prayer group, illustrates how survivalist knowledge incorporates the psychology of survival. Motsemme (2011) argues that survivalist knowledge takes different forms, such as Ukuhlonipha, roughly translated as 'respect', which particularly focuses on ways women support each other psychologically through embodied intergenerational rituals, whether through African traditions or the church, to enable healing from traumatic experiences. Recent scholarship of trauma affirms the importance of such embodied communal approaches to healing (Kolk, 2014). Ukuhlonipha's intergenerational embodied rituals often involve women-focused prayer groups (Motsemme, 2011). Despite the patriarchal nature of churches, these women's prayer groups are nevertheless worthy of feminist attention, as they enable and champion women's struggles, survival and resistance (Haddad, 2004).

The rural women's mobile diaries were filled with stories of death and the funerals that seemed to dominate the social calendar of rural life. Ma Sinazo had just lost her sister that Friday to a heart attack. When she took the emergency call from the hospital, she expected it to be news about her nephew, as he had been seriously ill for some time and her sister had been looking after him at home, only to find out it was her sister who had died. She attributed her sister's untimely death to the stresses of being both the breadwinner and the primary carer of the family in an era where there were no jobs for young people. She had left behind eight children.

> Ma Sinazo: It was a saddening story and not easy to take in. The kids were being supported by their mother and they are all not working. These kids are supposed to support us and we keep on hoping things will come right. We survive because of prayers.

It is important to note that such prayer groups are women exclusive and serve as a way of translating individual burdens into shared experiences of support. As Haddad (2004: 10) explains,

Prayer to God becomes a means through which women voice their burdens, away from sites of struggle, in their own safe space. It becomes an immediate link with the spiritual realm that enables them to see their lives from a different perspective as they unburden to God and to one another that which weighs heavily on their hearts. As they do this, women reach out with mutual care to one another and so 'become' the incarnate response to this pain. That which is expressed within the spiritual realm is manifested and dealt with in their human relationships with one another.

In this manner, the unbearable difficulties these women faced became bearable as they expressed their struggles and shared their burden with their sisters. As a committed member of the prayer group, Bulelwa was a very devout woman in her late thirties who recognized that rituals could help women deal with the trauma of a very difficult life through naming and recognizing shared difficulties, effectively making private suffering public and communal.

> Bulelwa: Everybody believes in something, some believe in the kraal that there are grandfathers there. That heals a person, it makes a person bold. I'm also talking about my own person, a person of Christianity ... there are times where you come together where you'd hear something comes from another person that would heal you. Maybe you're in a problem that person would say exactly what's inside of you then you'd be healed.

Bulelwa found the church an empowering force in her life that gave her the strength to get through an overwhelming schedule of daily tasks in the home she shared with her three boys. She listened to recordings of church sermons on her smartphone as she cooked porridge, did laundry, made steamed bread, swept the house and cooked soup. She was proud that she had created these recordings on her mobile phone herself. She explained that the congregation, consisting mostly of women, encouraged such uses of mobile technology.

> Bulelwa: Even the pastors use the phones in my church. While one preaches one pastor is recording also.

The level of women's digital ingenuity spread through the church was at times astounding, such as when Bulelwa revealed her digital frugality to exploit a hack in a VPN app that allowed her to briefly connect to the internet. Discourses of modernity associate technology with a notion of development propelling people to abandon traditional and religious village lifestyles for a secular individualistic modern lifestyle (Cornwall, 2007). Contrary to such assumptions, it was the church that was driving the adoption of advanced digital skills. Ma Rejoice had become an avid user of the King James Bible mobile app and particularly loved that she could now constantly share appropriate Bible verses with everyone. After downloading the Bible, she proceeded to download other books on gardening. Mobile technology has become so integral to the church service that Ma Sinazo shared her hopes of getting a smartphone soon, as she was the only one who didn't 'sing from her phone' in church.

In these spaces, it was the church that was the voice of civic care for others. For example, Liseko was inspired by that day's Bible study reading to share beetroot, spinach and carrots from her family's garden with various elderly people. Sustained by her close connections with her prayer group as a source of communal support, conversely, she found her interactions with the state through e-government frustrating and difficult, particularly since the time she had lost her phone in a taxi. An unemployed nursing assistant in her forties, Liseko had registered with the Department of Health to get mobile job alerts. Since there was no government office in the village, when she lost that phone and got a new phone and number, she had no way to change her login, and for a long time missed out on job alerts. Eventually, she realized she could associate an email address with this account and asked her sister, who has a smartphone, to use her email address. The irony is that the password-protected login designed to ensure Liseko's privacy now results in her constantly having to pester her sister to check for nursing job emails. As Dalvit (2024) argues from a study conducted in a similar rural village, e-government systems in South Africa are often profoundly alienating and disempowering. Ironically, one might consider the village women's prayer group as a more effective space of civic empowerment in comparison with e-government.

Nobomi and Nompumelelo ensure women's survival through privacy and safety work

The feminist slogan 'The personal is political' highlights the irony of trying to associate citizenship exclusively with public life. In this case study, we return to the Cape Town township to meet several young women trying to assert their privacy and safety rights in crowded conditions of urban life shaped by apartheid's spatial inequalities. Feminist scholars use the term 'safety work' to highlight women's invisible labour to maintain their safety and to underline the effort of such constant vigilance and draw connections with other forms of invisible and uncompensated labour expected from women (Vera-Gray and Kelly, 2020). Marwick (2023: 79) situates women's online safety and privacy in this tradition, arguing that women's work to maintain privacy can be seen as a form of safety work, and highlighting how it, like other forms of invisible work, is distributed unevenly by gender, race and class.

Nobomi was employed, living in a township house, sharing her small home with unemployed sisters and their children. She asserted her power as the breadwinner to define the rules about sharing the space. She did not allow her sisters to bring boyfriends home, explaining 'my house is not a *jent huis*' (whorehouse). Stepping back from the misogyny of the term *jent huis*, she clarified that she did not want to interfere with her sisters' freedom to experience intimate relationships, but to guard their home from the potential disruption of a man moving in

I'm not your mom I won't say you mustn't go and visit your boyfriend ... it's just that I don't want him to come inside the house.

Nobomi was engaged in gendered safety work as a practice of survival, warning her sisters of the importance of preserving their home as a refuge from GBV: 'don't bring your boyfriend to your home because one day your boyfriend would beat you, where are you going to run to?' Instead, Nobomi encouraged her sisters to create private spaces for their intimate relationships using mobile phones ('meet your boyfriend through the phone'). In the cities, safety work, such as using mobiles to set up travelling with friends or leaving smartphones at home to guard against having them stolen, was essential to guard against threats which are experienced intersectionally. The racial spatial inequalities of urban life under neoliberalism in Cape Town subjected women to many dangers, such as travelling long distances late at night for poorly paid shift work and evading workplace surveillance by their employers.

Nobomi's neighbour, Nompumelelo, engaged in another form of feminist labour, privacy work, to limit surveillance by her rural clan and to assert her power over her own sexual and romantic relationship. She was one of eight children and had a strained relationship with her mother, who lived in the Eastern Cape. As a young girl, Nompumelelo had been subjected to abuse by her mother after she was caught talking to a boy in public. 'She hit me there and then and threw away my Sony phone'. Even in adulthood, Nompumelelo's visits to the Eastern Cape usually ended in conflict. Arora points out that patriarchal vigilance around women's digital behaviour is often underpinned by hegemonic morality and religious institutions (Arora, 2019: 370).

On the previous day, Nompumelelo's mother had phoned to say she had heard of Nompumelelo's pregnancy. Nompumelelo rejected the overture, refusing her rural family any authority over her or her recent engagement. Traditionally, marriages are negotiated between extended families, but Nompumelelo had also opted out of this traditional practice. Instead, she announced her engagement by changing her Facebook status, thus informing more than two hundred Facebook friends, resulting in family members reprimanding her for not adhering to traditional protocols. Since then, she had decided to scale back her Facebook use, taking control of her privacy by selectively unfriending her rural clan and engaging in some self-censorship: 'I have decided that I will unfriend those that shouted at me'.

This disposition of scorning tradition to pursue romantic relationships might seem diametrically in opposition with the pious practices of Ukuhlonipha of the rural prayer group we encountered above. However, Motsemme (2011) argues that South African women's survivalist practices also include UkuPhanda or 'hustling', the ability to exploit sexual desire to manipulate men to ensure a woman's survival. Indeed, while rural women such as Ma Xoliswa may scorn the women who had left for the city to return wearing pants and asserting their superiority, this did not mean they were naïve to the patriarchal dynamics of traditional authority.

Conclusion

Our study has defined feminist digital citizenship through the practices of women who do not identify as feminist, who are only minimally online and who do not generally participate in what are considered civic activities. We nevertheless make the somewhat audacious claim that alongside what might be considered African elite spaces of social media feminism, the use of mobile technology for survival needs to be recognized as central to an everyday African feminist survivalist digital citizenship. It is a feminist digital citizenship that is primarily concerned with one fundamental human right: the right to life. Here, the right to life is not only something that can be taken away suddenly but is eroded bit by bit as the hardship of everyday life takes its toll. In these spaces, women tended to run households and needed to deal with the intersectional struggles of being women and Black and poor, where they were often breadwinners as well as primary caregivers. In our research site, institutions of government were largely absent or, if they were present, not very accessible. Women's struggles therefore largely focused on what Simone (2006) calls 'social and symbolic infrastructure', which emerges when ordinary people in the Global South create networks and routines that create stability and enable everyday life. Through constituting such social and symbolic infrastructure among themselves, they created spaces for living and embodied institutions of survivalist knowledge (Magoqwana, 2018). The stories that emerged in the mobile diaries challenge the stereotypes of African women as passive victims of their fate and instead reveal how these women take charge of their households and determinedly negotiate their family's survival with the help of their mobile phones while simultaneously crafting connectivity. Essential to such survivalist knowledge and practices is digital frugality, the knowledge and practices to sustain connectivity for less-connected people, which included an in-depth knowledge of service provider offers and practices of intermittently connecting and disconnecting, as Zintle's story illustrated. Under challenging circumstances, Zintle adeptly managed two network identities to maintain her privacy and maximize her accessibility, showing the importance of women's invisible labour and safety work. Similarly, Nobomi's use of online spaces for her sisters' intimate relationships with their boyfriends helped keep the peace and assert her power as breadwinner in a crowded household. Nompumelelo asserted her right to disconnect completely from rural relatives, claiming some privacy and control over her sexual relationships.

Ma Fezeka's story revealed how mobile connectivity underpins women's communicative labour, which holds together extended families and broader communities during times of disruptive civic action. Her story therefore showed how civic action may be gendered and that not being present in public spaces of protest does not necessarily show less commitment or active citizenship. Ma Xoliswa's story revealed the importance of being contactable by mobile phone in sustaining relationships of care, giving a technological character to the importance of responsiveness as a practice of care that indeed requires agency and knowledge even though this might not be obvious (Held, 2006). In reflecting on these two grandmothers' mobile diaries, we highlighted the need to place notions of care at the centre of citizenship (Lister, 2003) and argued for an African citizenship of care focused on the contribution of survivalist knowledge and practices of grandmothers (Motsemme, 2011; Magoqwane, 2018).

As the stories of the women of the prayer circle illustrate, such shared survivalist practices include the psychological care emanating from communal sharing of troubles or what Motsemme (2011) calls Ukuhlonipha. Such communal gatherings of care suggest religious women's organizations might be understood as contributing to civic life. Our mobile diaries illustrate how innovative uses of mobile technology emerge from the central role of spiritual organizations as spaces where women can support each other's psychological survival. In the story of the backyard dwellers, we show how navigating privacy and safety is similarly central to survival and incorporates various mobile practices that explicitly focus on sexuality and may therefore appear diametrically opposite to the pious practices of the prayer group. However, as Motsemme (2011) has indeed argued, the knowledges and practices of using women's sexual power in these spaces are also underpinned by the civic obligation of survival.

This chapter made visible women's digital survivalist practices as they incorporated the mobile phone into everyday activities of care. It also incorporated feminist invisible labours of care (Arora, 2019; Baym, 2015; Rakow, 1992; Vera-Gray and Kelly, 2020). We have therefore shown how the mobile phone is integral to African women's survivalist knowledge (Motsemme, 2011; Magoqwane, 2018). In particular, we have demonstrated how women's digital agency for connectivity, what we have called digital frugality, is part of African women's survivalist knowledge. We therefore extended such a focus on invisible human care to incorporate the invisible labour of digital frugality to create a 'lively infrastructure' (Amin, 2014) that enables the survival of communities and their digital connectivity.

Acknowledgements

We wish to acknowledge the central role of Indra De Lanerolle as principal investigator of the Izolo Mobile Diaries research project, who encouraged us to extend the findings of the study by reflecting on our respective field sites. We also acknowledge the crucial contribution of our research assistants, Samora Sekhukhune and Minah Radebe, whose astute knowledge of the social spaces in which we conducted the research, the fine nuances of isiXhosa and their deep ethical commitment to responsible research enabled the richness of the data and the integrity of the research. We also thank our translator, Khuselwa Tembani, who spent such care on conveying the meanings that emerged through the interviews. This chapter would not have been possible without the funding provided by Making All Voices Count to fund the Izolo Mobile Diaries study.

Note

1   The concept of being middle class is such a fluid concept in South Africa that both a South African who lives in a shack and one who owns a brand new Mercedes sedan could insist they are middle class (Alexander, 2013). Being middle class can alternatively be defined as being an average South African and ironically poor by global standards, or in terms of Global North notions of the middle class as having a professional job and therefore part of the South African elite.

Bibliography

Ajaja, S. (2023), 'Feminist Digital Citizenship in Nigeria', in T. Roberts and T. Bosch (eds), *Digital Citizenship in Africa: Technologies of Agency and Repression*, London, Bloomsbury Academic, 117–47.

Alexander, P. (2013), *Class in Soweto*, Scottsville, ZA: University of Kwazulu-Natal Press.

Amin, A. (2014), 'Lively Infrastructure', *Theory, Culture & Society*, 31 (7–8): 137–61. https://doi.org/10.1177/0263276414548490.

Arora, P. (2019), 'Decolonizing Privacy Studies', *Television & New Media*, 20 (4): 366–78. https://doi.org/10.1177/1527476418806092.

Baym, N. K. (2015), 'Connect With Your Audience! The Relational Labor of Connection', *The Communication Review*, 18 (1): 14–22. https://doi.org/10.1080/10714421.2015.996401.

Bosch, T. (2017), 'Twitter Activism and Youth in South Africa: The Case of #RhodesMustFall', *Information, Communication & Society*, 20 (2): 221–32. https://doi.org/10.1080/1369118X.2016.1162829.

Couldry, N. (2004), 'Theorising media as practice', *Social Semiotics*, 14 (2): 115–32. https://doi.org/10.1080/1035033042000238295.

Cornwall, A. (2007), 'Buzzwords and Fuzzwords: Deconstructing Development Discourse', *Development in Practice*, 17 (4–5): 471–84. https://doi.org/10.1080/09614520701469302.

Dalvit, L. (2024), 'The Hidden Colonialities of Mobile Communication: Phone Uses by Women in a South African Rural Community', in X. Pei, P. Malhotra and R. Ling (eds), *Women's Agency and Mobile Communication Under the Radar*, London and New York: Taylor & Francis.

de Lanerolle, I., Schoon, A. and Walton, M. (2020), 'Researching Mobile Phones in the Everyday Life of the "Less Connected": The Development of a New Diary Method', *African Journalism Studies*, 41 (4): 35–50. https://doi-org.ezproxy.uct.ac.za/10.1080/23743670.2020.1813785.

Dlamini, T. N. (2022), *The Experiences of Women Leaders in the #Fees Must Fall Movement: Narratives from the Universities of Johannesburg and Witwatersrand (2015-2016)*, [University of Johannesburg]. https://ujcontent.uj.ac.za/esploro/outputs/9925505107691.

Donner, J. (2015), *After Access: Inclusion, Development, and a More Mobile Internet*, Cambridge, MA: MIT Press.

Donner, J. and Gitau, S. (2009), *New Paths: Exploring Mobile-Centric Internet Use in South Africa*. 'Mobile 2.0: Beyond Voice?' Pre-conference workshop at the International Communication Association (ICA) Conference, Chicago, Ill., May 20.

Faith, B. (2018), 'Gender, Mobile, and Mobile Internet| Maintenance Affordances, Capabilities and Structural Inequalities: Mobile Phone Use by Low-Income Women', *Information Technologies & International Development*, 14 (0): Article 0.

Fester, G. (2005), 'Merely Mothers Perpetuating Patriarchy? Women's Grassroots Organizations in the Western Cape 1980 to 1990', in A. Gouws (ed.), *(Un)thinking Citizenship: Feminist Debates in Contemporary South Africa* (pp. 199–221), Farnham, UK: Ashgate Publishing.

Fischer, C. S. (1992), *America Calling: A Social History of the Telephone to 1940*, Berkeley, CA: University of California Press.

Gouws, A. (2016), 'Citizenship, Gender Equality and the Limits of Law Reform in South Africa', in E. N. Chow and E. Tastsoglou (eds), *Contours of Citizenship: Women, Diversity and Practices of Citizenship*, London and New York: Routledge, 723–67.

Gqola, P. D. (2007), 'How the "Cult of Femininity" and Violent Masculinities Support Endemic Gender Based Violence in Contemporary South Africa', *African Identities*, 5 (1): 111–24. https://doi.org/10.1080/14725840701253894.

Haddad, B. (2004), 'The Manyano Movement in South Africa: Site of Struggle, Survival, and Resistance', *Agenda*, 18 (61): 4–13. https://doi.org/10.1080/10130950.2004.9676032.

Held, V. (2006), *The Ethics of Care: Personal, Political, and Global*, New York: Oxford University Press.

Hernes, H. M. (1987), *Welfare State and Woman Power: Essays in State Feminism*, Oslo: Norwegian University Press.

Kolk, B. van der. (2014), *The Body Keeps the Score: Brain, Mind, and Body in the Healing of Trauma*, New York: Penguin.

Latzko-Toth, G., Millette, M. and Bonneau, C. (2016), 'Small Data, Thick Data: Thickening Strategies for Trace-Based Social Media Research', in L. Sloan and A. Quan-Haase (eds), *The Sage Handbook of Social Media Research Methods* (pp. 199–214), London: SAGE Publishing.

Leech, K. (2024), 'YOU ASKED: Has South Africa's public works programme created 14 million jobs in 20 years?', *Africa Check*, 0505-1313. https://africacheck.org/fact-checks/blog/you-asked-has-south-africas-public-works-programme-created-14-million-jobs-20.

Lister, R. (2003), *Citizenship: Feminist Perspectives*, New York: NYU Press.

Magoqwana, B. (2018), 'Repositioning uMakhulu as an Institution of Knowledge', in *Whose History Counts: Decolonising African Pre-Colonial Historiography*, Vol. 3, p. 75, Stellenbosch, ZA: African Sun Media. https://books.google.com/books?hl=en&lr=&id=QxJ9DwAAQBAJ&oi=fnd&pg=PA75&ots=7Wtm56Jzh1&sig=4PLhxhW9XZHesWjWNWXAJp971PQ.

Maluleke, G. (2024), 'Centering Orality in the Politics of Speaking Out Against Rape and Femicide On/Offline in South Africa', Unpublished Seminar Paper, May 6, Sociology Seminar Series, University of Cape Town.

Maluleke, G. and Moyer, E. (2020), '"We Have to Ask for Permission to Become": Young Women's Voices, Violence, and Mediated Space in South Africa', *Signs: Journal of Women in Culture and Society*, 45 (4): 871–902. https://doi.org/10.1086/707799.

Marwick, A. E. (2023), *The Private Is Political: Networked Privacy and Social Media*, New Haven, CT: Yale University Press.

Motsemme, N. (2011), *Lived and Embodied Suffering and Healing Amongst Mothers and Daughters in Chesterville Township, Kwazulu-Natal*, University of South Africa, Pretoria, http://hdl.handle.net/10500/5451.

Ojebode, A., Ojebuyi, B., Oladapo, O. and Oosterom, M. (2023), 'Ethno-Religious Citizenship in Nigeria: Ethno-Religious Faultlines', in T. Roberts and T. Bosch (eds), *Digital Citizenship in Africa: Technologies of Agency and Repression*, pp. 32–62, London: Bloomsbury Academic.

Pedersen, S. (1990), 'Gender, Welfare, and Citizenship in Britain during the Great War', *The American Historical Review*, 95 (4): 983–1006. https://doi.org/10.2307/2163475.

Rakow, L. F. (1992), *Gender on the Line: Women, the Telephone, and Community Life*, Urbana, IL: University of Illinois Press.

Richardson, C. and Skott-Myhre, H. A. (2012), 'Introduction', in C. Richardson and H. A. Skott-Myhre (eds), *Habitus of the Hood*, Bristol, UK: Intellect, 7–26.

Roberts, T. and Bosch, T. (2023), *Digital Citizenship in Africa: Technologies of Agency and Repression*, London: Bloomsbury Academic.

Robins, S., Cornwall, A. and von Lieres, B. (2008), 'Rethinking "Citizenship" in the Postcolony', *Third World Quarterly*, 29 (6), 1069–86. https://doi.org/10.1080/01436590802201048.

Ruddick, S. (1980), 'Maternal Thinking', *Feminist Studies*, 6 (2): 342–67. https://doi.org/10.2307/3177749.

Schoon, A. (2014), 'Digital Hustling: ICT Practices of Hip Hop Artists in Grahamstown', *Technoetic Arts*, 12 (2): 207–17. https://doi.org/10.1386/tear.12.2-3.207_1.

Schoon, A. and Strelitz, L. (2014), '(Im)mobile Phones: "Stuckness" and Mobile Phones in a Neighbourhood in a Small Town in South Africa', *Communicare*, 33 (2): 25–39.

Schoon, A., Mabweazara, H. M., Bosch, T. and Dugmore, H. (2020), 'Decolonising Digital Media Research Methods: Positioning African Digital Experiences as Epistemic Sites of Knowledge Production', *African Journalism Studies*, 41 (4): 1–15. https://doi-org.ezproxy.uct.ac.za/10.1080/23743670.2020.1865645.

Seekings, J. (2008), 'The Continuing Salience of Race: Discrimination and Diversity in South Africa', *Journal of Contemporary African Studies*, 26 (1): 1–25.

Seekings, J. (2015), *Continuity and Change in the South African Class Structure Since the End of Apartheid*. https://open.uct.ac.za/handle/11427/19180.

Sevenhuijsen, S., Bozalek, V., Gouws, A. and Minnaar-Mcdonald, M. (2003), 'South African Social Welfare Policy: An Analysis Using the Ethic of Care', *Critical Social Policy*, 23 (3): 299–321. https://doi.org/10.1177/02610183030233001.

Simone, A. (2006), 'Pirate Towns: Reworking Social and Symbolic Infrastructures in Johannesburg and Douala', *Urban Studies*, 43 (2): 357–70.

STATSSA (2015), *General Household Survey 2014* (Survey Results No. P0318), Statistics South Africa. http://www.statssa.gov.za/publications/P0318/P03182014.pdf.

Vera-Gray, F. and Kelly, L. (2020), 'Contested Gendered Space: Public Sexual Harassment and Women's Safety Work', *International Journal of Comparative and Applied Criminal Justice*, 44 (4): 265–75. https://doi.org/10.1080/01924036.2020.1732435.

Walton, M. (2014), 'Pavement Internet: Mobile Media Economies and Ecologies for Young People in South Africa', in G. Goggin and L. Hjorth (eds), *The Routledge Companion to Mobile Media*, pp. 450–461, London and New York: Routledge.

Walton, M. and Donner, J. (2012), *Public Access, Private Phone: The Interplay of Shared Access and the Mobile Internet for teenagers in Cape Town* (Global Impact Study Research Report Series), Cape Town: University of Cape Town. http://www.globalimpactstudy.org/wp-content/uploads/2012/11/Public-access-private-mobile-final.pdf.

Walton, M., Haßreiter, S. and Marsden, G. (2012), Degrees of Sharing: Proximate Media Sharing and Messaging by Young People in Khayelitsha. *MobileHCI*, 12, 403–12.

Chapter 6

#GUINEENNEDU21ESIECLE AND THE RADICAL POTENTIAL OF FEMINIST ACTIVISM IN CONTEMPORARY GUINEA

Clovis Bergère

This chapter examines the feminist campaign Guinéenne Du 21e Siècle [twenty-first century Guinean Woman], sometimes abbreviated to Gdu21es, a recent example of digital citizenship that has been active since launching in 2016 in the Republic of Guinea, a small coastal Francophone West African country with a population currently estimated at around 13 million. The Guinéenne Du 21e Siècle campaign launched in March 2016 to coincide with that year's International Women's Day on 8 March and initially centred on the hashtag #guineennedu21esiecle [#21stcenturyguineanwoman]. It was spearheaded by an initial core group of around twelve young Guinean women web-activists working under the auspices of the eponymous collective Guinéenne Du 21e Siècle. The impulse behind the campaign was in part triggered by a frustration with the way in which International Women's Day was typically celebrated in Conakry, Guinea's capital, in the early 2000s. As Diérétou Diallo, the young Guinean web-activist and blogger behind the campaign, explained in an interview with the now-defunct Guinean online news site guineeinfos.org on 3 March 2016,

> *Un matin, c'était il y a deux jours, je me suis réveillée et j'ai réalisé que dans une semaine chrono, nous serions le 8 Mars 2016. Je me suis alors posée la question « Comment fête-t-on habituellement le 8 mars à Conakry ? », la réponse est simple, presque banale tant on peut la deviner: habituellement, les guinéennes se dotent de pagnes imprimés à l'effigie de grandes dames africaines ou de personnalités politiques puis vont au Palais du Peuple, se livrent à quelques danses rituelles, à quelques discours ici et là puis à la fin vers 19h, tout le monde plie bagage et rentre chez soi. Fini la fête de la femme. Cette année, j'ai voulu bouleverser les codes, apporter une nouvelle donne et pourquoi pas en faire une habitude ? J'ai lancé une campagne numérique sur 4 réseaux: Twitter, Facebook, Instagram, et Youtube.*
>
> [One morning – that was two days ago – I woke up and realized that in exactly one week, it will be March 8th 2016. I then asked myself: 'how is March 8th

habitually celebrated in Conakry?', the answer is simple, almost clichéd given how predictable it is: usually, Guinean women adorn their African cloth dresses printed with the portraits of illustrious African women or politicians and head to the Palais du Peuple, do a few traditional dances, give a few speeches here and there until about 7:00 pm, and everybody packs up and goes home. Women's Day celebrations are done and over. This year, I wanted these to be different, to bring a new perspective and why not turn that into a new habit? I launched a digital campaign across four platforms: Twitter, Facebook, Instagram, and YouTube.]

Since 2016, in addition to the initial #guineennedu21esiecle hashtag campaign, the initiative has included several web-based activities, including a blog, a digital 'blackout' campaign to protest the mistreatment of the young M'mah Sylla who died as a result of sexual violence and medical mistreatment, and several awareness campaigns on topics such as female genital mutilation, the conditions of Guinean women working in Kuwait and breast cancer, for instance. The collective has also organized a number of on-the-ground initiatives including training on digital citizenship for women, as well as charity runs.

Gdu21es is part of a broader and growing movement spearheaded by a new generation of Guineans – mostly urban, educated and middle class – to mobilize the organizing power of social media, Twitter and Facebook in particular, to bring issues and issue-based activism to the fore of the Guinean public sphere (Bergère, 2019). It does stand out in the history of digital citizenship in Guinea for several reasons that make it a useful entry point for a broader reflection on contemporary feminist citizenship in Francophone West Africa.

First, Gdu21es is led by a collective of young women activists when most citizenship campaigns in West Africa have tended to be led by men.[1] Second, the campaign is also notable for attracting significant attention at the time, both locally in Guinea, where it generated hundreds of tweets and vivid discussions online, and internationally. Regional and international media in Senegal, France and Spain, for instance, featured articles about the campaign, including interviews with the activists behind it. Although it launched in 2016, the campaign also stands out in the history of digital citizenship, as it has continued to exist over time and remains active at the time of writing in 2024, whereas most online rights-claiming issue-based campaigns in Guinea have tended to be limited in time. Third, the campaign was born as a reaction to prior feminist activist practices, in particular to the way International Women's Day is typically celebrated in Guinea, as noted in the guineeinfos.org interview excerpt above. As such, the campaign offers a unique opportunity to examine the changing landscape and intersections of feminist activism and digital citizenship in contemporary Guinea and, perhaps, more widely across Francophone West Africa.

Specifically, this chapter is concerned with three interrelated questions: 1) If Gdu21es is widely understood as a feminist digital citizenship campaign, which is how the collective behind it describes it, what can it tell us about what a 'feminist' approach to digital citizenship might entail? 2) Furthermore, how does Gdu21es relate to the long history of feminist activism in Guinea, where, as in much of West Africa, women have played a central role in citizenship, activism and radical politics, pre- and post-independence, from the 1940s onwards? 3) Finally, what specific, embodied, historicized and emplaced expressions of citizenship and rights claims does the campaign Gdu21es foreground, contest and contend with?

To answer these questions, the chapter draws on a multi-pronged approach. This includes data collected as part of a digital and grounded ethnographic research project that I conducted between 2014 and 2017 as part of my doctoral dissertation project. In addition to this ethnographic data, I focus much of the analysis on a small corpus of around eighty-five tweets, sixty Facebook posts, content from the blog Guinéenne Du 21e Siècle and news articles related to the campaign in both local Guinean outlets and regional and international press, such as France 24, BBC Africa and *El País*.

Given the importance of context-specificity and the need for research on the digital that is grounded in local context, I begin the chapter by briefly retracing key moments in the recent history of feminist activism in Guinea. This is inspired by Jack and Avle's (2021) feminist geopolitics of technology framework. This framework emphasizes knowledge of the digital that is historically informed, meaning that it 'considers histories of colonialism and racial capitalism in analyzing contemporary power relations' (ibid: 2021: 1). A feminist geopolitics of technology framework also draws on intersectional feminist literature to encourage 'attention to the lived experiences of marginalized populations, including those in the Global South,' thereby helping in 'analyzing various lines of affinity and dis/empowerment within and beyond the unit of the nation-state' (ibid: 2021: 1). This first section starts with a brief account of women's central activism within the Rassemblement Démocratique Africain (RDA) which led Guinea to independence in 1958, their roles as both forebearers and chief contesters of the First Republic's revolutionary politics, and their more recent activities in the 2006/7 general strikes and the 2014 'Market Tax' protests in Labé, all drawing on a long history of grassroots organizing and feminist defiance in Guinea, and West Africa more generally (Schmidt, 2005; Pauthier, 2007; Diabaté, 2020).

A second section, then, provides more details on Gdu21es, focusing most particularly on the initial hashtag campaign #guineennedu21esiecle itself. This includes describing in detail the content produced as part of the campaign and the strategies deployed to make its specific claims as well as drawing on interviews with Guinean web-activists, including several closely associated with the campaigns, to better situate them within the broader landscape of digital citizenship activism in the country. The third section follows Jack and Avle's invitation to draw on intersectional feminist scholarship by examining how the notions of citizenship and related concepts of public sphere and belonging have been discussed in African political philosophy and media studies, drawing most specifically on the writings of Aminata Diaw, Nadia Yalla Kisukidi, Wendy Willems, Omotayo Jolaosho and Hourya Bentouhami, among others. This is, then, briefly related to recent discussions on the digital and the digital politics of matter and transparency within Black and African studies, in the works of Florence Madenga and Armond Towns, for instance.

The next section analyses the digital content produced as part of the #guineennedu21esiecle campaign, drawing on the historical data and theoretical debates presented in sections one and two to argue that 1) the digital feminist citizenship campaign #guineennedu21esiecle not only showcases the diversity and at times contrary nature of feminist activist practices but also speaks to the tensions within what Jolaosho describes as 'awkward activism'; 2) digital citizenship campaigns such as Guinéenne Du 21e Siècle are extensively informed by long-standing feminist practices, including communication practices, embedded in local communities and geographies that help explain their success and reach; and 3) rather than sitting at a rupture with the past, recent digital citizenship campaigns such as Gdu21es inhabit complex and contested historical junctures, made up of past failures and successes. As such, they continue to contain some of the radical potential of feminist politics as dreamt in the independence period, even if they purport to disrupt them. This is contrary to the framing of campaigns such as Gdu21es within neoliberal imaginations of the digital, as exemplified in framings of Gdu21es within the international press, for instance.

From 'dangerous amazons' to 'chief sloganeers': Feminist mobilizations in Guinea since 1945

The historiography of digital campaigning in Guinea, as elsewhere, has tended to be told in male-centric ways that obscure the active participation of women in struggles for liberation across the world (Beoku-Betts and Ampofo, 2021: 10). As such, it reproduces the historical erasure of women's roles in activist campaigns for more inclusive rights, citizenship and just lives. To fully grasp the cultural and historical significance of the recent digital campaign #guineennedu21esiecle, it is, therefore, necessary to situate it within the long and rich history of feminist organizing and struggles in Guinea on which it directly drew, both explicitly and implicitly. The history cursorily retraced here is both unique to Guinea and also representative of many other struggles for inclusion, equality and emancipation that have been led by women across the world, and in the Global South most particularly (see Beoku-Betts and Ampofo, 2021, for instance).

1945–1958: Women take the lead

To say that women played a central role in the fight for independence from French colonization is an understatement. Their activities as part of the Parti Démocratique Guinéen (PDG), the local branch of the pan-African RDA, which led the fight for independence in Guinea, were crucial to the party's success. As PDG-RDA activist Aissatou N'Daye has claimed, 'Women brought about independence. It was really the women' (cited in Schmidt, 2005: 114). As exemplified by early women activists, such as Jeanne Martin, Nankoria Kourouma and Sarata Diané, women's initial involvement in the struggle for independence as part of the PDG-RDA was limited to a small group of elite women, typically Western-educated and often politicized through their husbands' own leading roles within the party. The general strike of 1953, however, marked a watershed moment when a large coalition of women transcending class, ethnic, religious and caste allegiances entered the forefront of Guinean politics. For the first time, a sweeping coalition of non-elite women mobilized *en masse* in support of the general strike and nationalist movement more broadly. Their involvement in the strike was wide-ranging and brought about new tactics of protest. Women marketeers boycotted European retailers. At a pivotal meeting in the cinema Rialto in Conakry, when strikers were trying to decide whether to extend the strike or not, a committee of around five hundred women spoke out and enjoined their hesitant male counterparts to continue the general strike.

Perhaps most saliently, through their active involvement in the 1953 general strike, which lasted a sweeping seventy-two days, Guinean women clearly demonstrated their enormous mobilization power, something that has continued to mark Guinean political life since. Through long-standing practices, including traditional mutual aid associations and extensive networks across everyday spaces, such as markets, public fountains and taxi stations, women brought new ways to circulate information and thus reach vast segments of the population. This deeply grounded, highly efficient and nimble communication network, sometimes referred to as 'bush telegraph', allowed rural, poor and non-elite women who had previously been excluded from revolutionary politics to be brought to the fore and mobilized. According to several Guinean scholars present at the time, women drew on their historical roles as storytellers or singers to create messages that appealed to as wide a segment of the population as possible, thereby becoming known informally as the PDG-RDA's best 'sloganeers' (Schmidt, 2005: 127). As a result of their active involvement in nationalist politics, many women were arrested and imprisoned by colonial authorities; chief among them were prominent activists such as Nabya Haidara and Mafory Bangoura, widely known in Guinea as the 'dangerous amazon of the PDG' (ibid.).

1958–1984: Militancy and emancipation under the revolution

It is worth noting that women's abilities to create songs and slogans that appeal to a broad swath of the Guinean population were strongly influenced by long-standing histories of women's militancy, including pre-colonial forms of activism that made use of physical interventions, sex strikes and at times violent means of action when necessary (Rivière, 1971; Diabaté, 2020; Joalosho, 2019). As a result, women's collective actions, slogans or songs could be violent, rude or sexually suggestive. Women's ability to mobilize these potent discursive registers was in part what made them appealing as 'troupes de choc', or shock troupes, for the PDG-RDA. After Guinea gained independence from France in 1958, and the PDG assumed power in the newly independent Guinea, women's role – alongside youth – at the forefront of revolutionary politics had been cemented. The PDG, and Sékou Touré chiefly within it, recognized the need to secure women's support. As a result, the party and president espoused an official rhetoric in support of women's rights and emancipation, embracing at times radical notions of equality and foregrounding topics reflecting concerns close to women's interests, such as education, employment and political participation. They also institutionalized women's organizational power, forming eight thousand special committees, each headed by an executive committee made up of at least thirteen women, resulting in at least 104,000 women in charge across the country. These facts were often repeated to Guinean publics during the revolutionary era through widely distributed pamphlets such as the 1977 *La Femme dans la Société*, a required reading as part of the ideological training seminar of that year's Guinean university graduating class, which reiterated these statistics, noting, for instance, that 'these numbers underscore the most notable fact that out of 35 women in Guinea, there is at least one exercising functions of public responsibility' (Touré, 1977: 69). Centralized and highly hierarchical, these women's committees played a key role in spreading locally the party's various orders and policy briefs. However, the revolution's discourse on the emancipation of women is clearly marked by a profound ambiguity: while women were officially elevated to the forefront of revolutionary politics, their roles as women were clearly defined by the party, and any deviation or challenge to the official line was chiefly reprimanded.

The protests of 1977, sometimes referred to as the 'women's revolt', mark another key turning point in the history of emancipatory struggles in Guinea and are widely noted as the largest open contestation to the Sékou Touré regime. Faced with an increasingly authoritarian revolutionary regime, mired by a violently paranoid president living in permanent fear of a local or foreign plot, women marketeers across Guinea took to the streets in open opposition to the regime. In June 1977, exasperated by constant harassment under the economic policy, women selling in the market in N'Zérékoré, the country's second-largest city, protested so vehemently that the local militia had no choice but to turn back after leaving two dead behind them (Pauthier, 2007: 231). A few months later, in August 1977, a similar altercation between a women marketeer and a member of the economic police at the M'Balia market in Conakry resulted in a collective action by the market women, who decided to march towards the Palais de la Présidence, the president's residential palace. There, joined by vendors from other markets, as well as youths, several thousands gathered in protests. The women drew on their long-standing protest repertoire, some singing chants such as 'vingt ans de crimes, c'est assez, tu dois t'en aller' [twenty years of crime, that's enough, you must go] and hurling insults directly at Sékou Touré. According to several accounts, the protesters were so loud that Sékou Touré could not be heard speaking, some claiming he even had to flee (Ammann, 2020: 37). The next day, thousands came back in protests, many wearing red in open defiance to Sékou Touré's order to wear white to signal commitment to the revolution. Clashes continued for several weeks until early September, once again cementing Guinean women's abilities to mobilize huge segments of the Guinean population in the struggle for popular emancipation.

Since 1984: Liberalization and the changing landscape of contestation in Guinea

Following the death of Sékou Touré in 1984 and the fall of the First Republic, women have continued to occupy a central role in Guinea political life, actively contributing to emancipatory struggles. Some of the forms and contours of feminist citizenship and campaigning post-1984 are in direct continuation of women's involvement in nationalist politics both pre- and post-independence. As Pauthier (2007: 233) has noted, while there needs to be more systematic examination of the links between pre- and post-revolutionary era feminist struggles, there seem to be clear lines of continuities in both content and practice of mobilization as women reinvest their experiences from both the fight for independence and within the structures of the PDG to shape their involvements with new NGOs or civil society organizations. Some of the leading figures of the trade union and women's movements of the post-1984 period, such as Hadja Rabiatou Sérah Diallo, who headed the country's most powerful workers' union between 2000 and 2011, began their careers during the First Republic years. During popular protests such as the 2006 and 2007 general strikes, the largest protests since the 1953 strike discussed above, or more recently the 2014 market tax protests in Labé – a major city in central Guinea – when women market vendors refused to pay the market and hurled insults at municipal officers, women protesters drew directly on the tactics forged in the fight for national liberation. The years since the fall of the socialist republic have been marked by the rapid imposition of market economics and neoliberal orthodoxies, including devastating structural adjustment policies, enforced retreat of the state and the privatization of social care, with a growing role for civil society organizations and NGOs. This has deeply shaped new forms of feminist struggles, now much more influenced by global agendas and individualized tactics. These have been marked by efforts to distinguish themselves from the revolutionary politics of the post-independence era while simultaneously grappling with the implications of the neoliberal marketization for advancing feminist agendas and emancipatory politics in Guinea. Yet, as Pauthier reminds us: 'The demonstrations of 1977 remain in people's minds as proof of women's abilities to confront law enforcement forces and to impose the voice of the people to those in charge' (2007: 237).

News reporting on recent digital citizenship campaigns, such as #guineennedu21esiecle, tend to emphasize the newness of such campaigns. According to *El País*, for instance, such campaigns allow Guinea, a country that as the Spanish newspaper explained somewhat dismissively, 'is not exactly a technological center', to enter digital modernity. It, however, is important to consider the opposite: beyond ruptures, how much do recent digital campaigns in Guinea, and West Africa more broadly, carry forward rich histories of feminist activism and citizenship that, in fact, shape what feminist citizenship might look like in the digital age? Obscuring these lines of historical continuation fails to acknowledge the multiple contributions of locales widely understood as on the margins of digital innovation to digital life, and citizenship more broadly.

Description of #guineennedu21esiecle

Planned to coincide with International Women's Day on 8 March 2016, as noted above, the launch of the digital campaign #guineennedu21esiecle was announced about a week earlier, in early March. In a detailed Facebook post, the collective Guinéenne du 21e Siècle explained both the rationale behind the campaign and how to participate. As the post described, to champion feminist causes in general and Guinean feminist causes in particular, internet users were invited to 'come, tell, and evoke what a modern Guinean woman represents for them, what they would like her to be, what conditions should be made available to her, the rights she lacks, and solutions to get there'. As the collective explained, the campaign would use four platforms: Facebook, Twitter (now X), Instagram and YouTube. The initial Facebook post launching the campaign then provided technical details on how to participate. Specifically, it invited those without a Twitter (now X) account to sign up for one, as 'regardless of the campaign itself, having one is useful professionally in the digital era'. The announcement also specified the length of the videos to be posted on YouTube, which should be longer than two minutes. Importantly, the campaign organizers asked participants to include in all campaign-related posts a photograph of themselves holding an A4 sheet of paper with the hashtag #guineennedu21esiecle written on it, preferably using a black or blue marker to ensure that the hashtag is clearly visible. This was a key visual element of the campaign and resulted in hundreds of posts on social media, Twitter (now X) and Facebook in particular, including photographs of participants holding a handwritten A4 sign as specified. Although a fair number of men, Guinean men in particular, participated in the campaign in March 2016, the vast majority of posts came from Guinean women, despite the fact that the organizers made it clear that the campaign was by no means restricted to women participants. In fact, the post clearly explained that men as well as non-Guineans were welcome to take part. The photographs were accompanied by typically short commentary. I provide below a sample of commentary that accompanied the photographs responding to the campaign's invitation to express what constitutes for the participants the ideal modern twenty-first-century Guinean woman. Some of these are taken from Twitter (now X) or Facebook posts, and some from short videos posted on YouTube. The platform used is specified each time, along with English translation:

> La #guineennedu21esiecle est une guerrière et ne peut être assimiler à la faiblesse
> [The #21stcenturyguineanwoman is a warrior and cannot be associated with weakness – Twitter (now X)].
> Ce sont des femmes qui font la différence dans l'ombre #guineennedu21esiecle

[These are women who make a difference in the shadows #21stcenturyguineanwoman – Twitter (now X)].

La #guineennedu21esiecle c'est celle qui pousse ses pairs vers le haut, ensemble nous irons plus loin [The #21stcenturyguineanwoman is the one who pushes her peers upwards, together we will go further – Twitter (now X)].

On est des warriors, des gladiatrices [we are warriors, gladiators – YouTube].

Elle s'assume financièrement [She is financially independent – YouTube].
Elle dit non aux mutilations genitales feminines [She says no to female genital mutilation – Twitter (now X) and YouTube].
Elle participate aux prises de decision [She participates in decision-making – YouTube].

Elle ne se dépigmente pas [She does not bleach her skin – Twitter (now X)].

Les Guinéennes Du 21ᵉ Siècle sont des actrices incontournables des sociétés sociales et économiques actuelles, il faut saluer leur efforts répétés pour l'émergence de nos nations, qu'elles soit travailleuses, cadres, étudiantes, fonctionnaires, commerçantes, mères au foyer [21st century Guinean women are major actors of our current social and economic societies, we have to acknowledge their continued efforts for the emergence of our nations, whether they are workers, managers, students, public employees, shopkeepers, staying at home – YouTube].

#guineennedu21esiecle est intelligente, persévérante, courageuse, instruite [21stcenturyguineanwoman is intelligent, persevering, courageous, educated – YouTube].

Elle dit non aux violences faites aux femmes [She says no to violence against women – Twitter (now X) and YouTube].

While the examples shared here only represent a very small portion of the content produced as part of the hashtag campaign, they illustrate some of the major themes covered in the user-generated response to the prompt. The campaign was widely celebrated as a success and received a great deal of media attention. Most major news outlets in Guinea, including Guinéenews and guineeinfos.org, as well as several major international media organizations, such as France24, *Jeune Afrique* and BBC Africa, published features on the campaigns. The Spanish *El País*, for instance, published an article that described the campaign as follows:

A group of young Guinean women have taken over social networking sites and managed to draw the attention of international media. They did it using Information and Communication Technologies (ICTs) and social networking sites to put together an operation without over-sentimentalizing the issue. They are calling for a new Guinea, a modern Guinea, and a constructive spirit. These are women, youths and bloggers, and their weapons are creativity and innovation.

(*El País*, 4/23/2016, translation is mine)

In many ways, the trajectory of the #guineennedu21esiecle hashtag and campaign, briefly outlined here, as well as its celebration in various international media as an example of a new expression of citizenship on the part of a new kind of Guinean women, young, 'modern', creative and media-savvy, aligns closely with broader narratives about the political potential of digital technologies for reimagining citizenship. One the one hand, they reflect a widespread celebration of digital 'technologies of extension' (Nyamnjoh, 2022) as liberatory, enabling citizens to rid themselves of prior markers of class, gender and race, facilitating forms of organizing previously unthinkable in real life. On the other hand, by erasing the historical contexts of these digitally mediated citizenship campaigns, dominant accounts of digital citizenship campaigns, as exemplified by the *El País* article briefly quoted above, tend to offer a narrative that rejects 'over-sentimentalizing the issue' and is stripped of its radical potential, including its feminist dimensions and inscription within a long history of radical feminist citizenship in Guinea. I return to this later in the chapter.

Embodied and contested notions of citizenship

One of the many contributions of feminist scholars to our understanding of citizenship and the related notions of public sphere and belonging has been to demonstrate that Western modernity has not only promoted deeply gendered and racialized notions of political participation but also worked to legitimize such notions primarily through erasure, silencing and invisibilization, rather than (and of course at times in addition to) direct confrontation. Nadia Yala Kisukidi, for instance, opens a chapter on the practice of decolonizing philosophy by reminding us of what she calls a '*vieillerie*' ['old thing' or 'old idea' in English], something that we shouldn't have to be reminded of anymore: '*la philosophie possède un corps, une couleur, un lieu*' [philosophy possesses a body, a color, a place] (2017: 53). In other words, concepts such as citizenship do not happen in a vacuum; even when they claim to apply universally, indifferently of bodies, color, places, they in fact reflect the bodies and places that birthed them.

In the context of African political philosophy, feminist scholar Aminata Diaw, for instance, shows that modern conceptions of citizenship derived from the universal ideals of the Enlightenment reflect a deeply gendered – and racialized – hierarchization of political participation, all the while acting in the name of supposedly inalienable individual freedom and equality. As she explains,

> [The] exclusion of women from the realm of citizenship seems to be a consistent element of political philosophy that, drawing on metaphysics from Plato, Aristotle, then Descartes, and beyond, has proceeded by creating hierarchies based on duality: body/mind, body/soul, matter/form, sensible/intelligible,

activity/passivity, etc. This exclusion has formed the basis of modern notions of citizenship, even though one of the major contribution of modern political philosophy remains the unequivocal affirmation of the principle of natural equality amongst all humans. Of prime importance here is the fact that modernity has constructed the notion of citizenship on the basis of the exclusion of women and, therefore, on the crystallization of the persistent boundary that delineates the public and private, the domestic sphere and public space, the domestic government of the family and the government of the city, or *polis*.

(Diaw, 2009: 50)

Offering a critical reading of the writings of Rousseau, whose political philosophy looms large in modern conceptions of citizenship, Diaw demonstrates how seemingly radical affirmations of equality among citizens, for Rousseau – as for most thinkers of the Enlightenment – in fact rest on a fundamental distinction between the public sphere of the *polis*, organized through human design, and the private sphere, which responds to what Rousseau understands as the principles of nature. For Rousseau, in the domestic sphere, women are subordinate to men. This separation of the public and private excludes women from full and active participation in the life of the *polis*, rendering them effectively silent and invisible. As Diaw explains, drawing on the work of Luce Irigaray and Carole Paterman, this silencing, erasure and absence of women from the realm of citizenship is linked to the fact that in modernity the 'fundamental model for the human subject remains unchanged: One, unique, solitary, and historically male, [...] Western, adult, reasonable, competent' (Irigaray, 2002). It has also meant that women's active participation in the affairs of the *polis* has historically been equated with a challenge to public order, a 'public disorder' (Diaw, 2009: 54).

Similarly, in her 2023 article 'The Reproduction of Canonical Silences', Wendy Willems re-reads Habermas's highly influential work on the public sphere (2023). As she demonstrates, for Habermas, the burgeoning coffee house culture, supported by the coeval emergence of newspapers, journals and periodicals in seventeenth- and eighteenth-century England is fundamentally constitutive of a new bourgeois public sphere 'where private people come together as public and discuss matters of common concern' (Habermas, 1991: 27). Conspicuously absent from Habermas' historical account of the emergence of this new European modern liberal sensibility and citizenry is the triangular economy centred around the transatlantic slave trade, plantation economy, colonial conquest and racial hierarchies that made the very existence of the coffee houses celebrated by Habermas possible in the first place. Echoing Diaw's argument about the erasure of women from modern conceptions of citizenship, Willems shows how conceptions of the public sphere central to modern political philosophy have equally silenced the racialized bodies and colonial hierarchies on which they rest.

African and Afro-diasporic feminist scholars such as Nadia Yala Kisukidi, Aminata Diaw, Abosede George and Omotayo Jolaosho have sought to counter abstract universalist conceptions of political participation and activism that reproduce the silences underpinning their constitution with plural conceptions of citizenship as necessarily gendered, racialized, embodied and historicized practices. As Tony Roberts and Tanja Bosch noted in the introduction to *Digital Citizenship in Africa*:

> Citizenship does not occur in a vacuum; it is expressed in spaces and places (Jones and Gaventa 2002), and the specific historic, cultural and power relationships of those spaces inevitably shape the temporal and situated meaning and practices of citizenship in those places. This makes the situated study of citizenship in particular geographies and within specific groups essential to a full understanding of digital citizenship in Africa.
>
> (2023: 4)

This argument is echoed in Omotayo Jolaosho's work on coalitional activism in South Africa. Drawing on feminist critiques that posit the situatedness of knowledge (see Kisukidi, for instance), Jolaosho noted the dissonance that often exists between 'stated commitments to gender equity and activist practices' (2018: 425). She proposes a framework for analyzing feminist activism that distinguishes feminist approaches from dominant coalitional practices, noting three points of dissonance, including: 1) the cultivation of women's autonomous perceptions, 2) the primacy of embodied experience, and 3) an emphasis on voice and dialogue (2018: 426). What emerges in Jolaosho's examination of feminist activist practices in South Africa are fundamentally embodied and active conceptions of political participation that reveal the foundational erasure and silencing of women's voices within activist circles that purport to act in the defense of gender equity.

Feminist writing on political concepts such as citizenship, activism, and politics allows us to foreground three specific aspects of political life that are particularly relevant to how we understand digital citizenship. The first is the fact that bodies matter. This is particularly important to remember in relation to the digital, which has sometimes been understood as disembodied. Feminist scholars' such as Kisukidi, Diaw, Jolaosho's insistence on the need to account for the erasure of gendered, racialized, classed and otherwise intersectionally invisibilized bodies echoes recent scholarship on the digital, including the work of Armond Towns on race and digital technologies, which has sought to deconstruct the dominant understandings of the digital that emphasize transparency and scientific neutrality. As he notes:

> In the wake of Alan Turing's machine—one of the central gatherers of information toward breaking the Nazi's Enigma code during World War II—the digital is increasingly equated with the promised delivery of transparency by the end of the twentieth century. This scientific approach to communication and information bled outside of computational calculations, giving birth to arguments that human communication, too, could be quantified and scientifically analysed—sparking the mid-twentieth century development of social scientific elements of the discipline of communication. The transparency of the digital media economy is based on some, not all, humans.
>
> (2021: 124)

At stake, for Towns, is the fact that within this discourse, Black and brown bodies, stripped of their humanity, cannot be imagined beyond mere matter that is inanimate, passive rather than active. Furthermore, the discourse of digital transparency also corresponds to the emergence of the related discourse of the democratic potential of digital circulation since, as Towns notes, in the digital sphere, we can theoretically 'all buy, sell, and make content, rather than just have it made for us' (2021 130). While based primarily in the United States, such arguments against liberal notions of transparency are echoed in work on digital technologies in Africa (see Madenga, 2019 and Nyamnjoh, 2022). If digital technologies have in Africa provided ordinary people with the opportunity to carry out their struggle against authoritarian regimes, those technologies have also negatively affected the very democratic potential they purport to promote as they continue to 'impose hierarchies of visibility narrowly configured to satisfy the logic and desire for profit' (Roberts and Bosch, 2023: xvii). For Towns, such discourses 'conflate democracy with the market' and 'Black and brown revolution [...] with the enrichment of a small Black and brown petit bourgeoisie' rather than radical political struggle (Towns, 2021: 130). Inspired by Diaw, Kisukidi, Willems and Jolaosho's feminist critiques of citizenship; Nyamnjoh's notion of flexible citizenship; and Towns's critique of digital transparency, this chapter seeks to historicize recent digital citizenship campaigns in Guinea, using #guineennedu21esiecle as a case study on women's long-standing radical politics and citizenship mobilization, which foreground understandings of citizenship as active, situated and flexible rather than bounded and transparent.

Analysis

The initial Gdu21es campaign, centred on the hashtag #guineennedu21esiecle, its inception, and the content generated by users responding to its invitation to 'come, tell, and evoke what a modern Guinean woman represents for them, what they would like her to be, what conditions should be made available to her, the rights she lacks, and solutions to get there' speak to the diversity of feminist citizenship practices as well as the tensions that arise from the confrontation of multiple epistemologies which they represent. This relates directly to what Jolaosho describes in her work on coalitional feminist politics in South Africa as 'awkward activisms', which she insists require 'creating new activist practices rather than duplicating problematic conventions even if these had more legitimacy' (2018: 444). As noted in the description of #guineennedu21esiecle above, the Gdu21es collective was focused on challenging the status quo and creating new ways of organizing around International Women's Day in Guinea.

For Jolaosho, feminist activist practices are best understood as 'awkward', building on the work of Marilyn Strathern on awkwardness, because they are necessarily composed of a plurality of epistemological horizons that often come to highlight the dissonance and even discordance between various agendas, practices and feminist intervention. The brief list of examples of user-generated content shared above speaks to this plurality and the tensions arising from it. While some insisted on the twenty-first-century Guinean woman's need for independence, others reiterated the need for more traditional gender roles within the family unit, defined as a union between a man and woman, for instance. Other tensions arose with regards to class, with some for the commentary equating being 'modern' with professional success in fields such as law, engineering or public service that reproduces a middle-class ideal often conditioned on fair access to quality education that is not realizable for the vast majority of Guinean women today. Others insisted that street vendors, traditional performers and those staying at home were equally part of what makes the ideal twenty-first-century woman. Some content focused on matters such as skin lightening, intimate relations or GBV that sit awkwardly and sometimes in tension with dominant activist agendas in Guinea, often led by men.

What is more, for Jolaosho, a feminist activist practice brings to the fore an ethos that is rooted in embodied experience. This works to counter the systemic erasures of women and racialized bodies from dominant notions of citizenship originating in Western modernity and colonial imagination, as critiqued by Diaw, Willems, Kisukidi and others. The insistence that participants in the digital campaign #guineennedu21esiecle include a photograph of themselves holding an A4 sheet of paper with the hashtag written on it is here therefore significant. It enabled the 'body, color, and place' – to borrow Kisukidi's words quoted above – to be visible, accounted for and fully acknowledged. A slideshow of all the pictures emerging from the campaign and posted in 2016 on the Guinéenne Du 21e Siècle blog shortly after the campaign ended provided a powerful counter to both understandings of citizenship as disembodied and digital technologies as transparent. It provided a visual archive of what feminist notions of citizenship stemming from West Africa might look and feel like while also foregrounding digitally mediated Black bodies as active and agentic rather than mere matter (Towns, 2022).

In her essay 'Notes pour Feminisme Marron' (Bentouhami, 2019) on what revolutionary potential for a feminist politics might the history of women's role in marronage, the process of extracting oneself from slavery, hold, Hourya Bentouhami reminds us that central to women's ability to escape slavers' control and pursuit was marron women's 'maîtrise parfaite du territoire' [perfect mastery of the territory] (2019: 39). As noted above, historically in Guinea, it was also women's mastery of the territory, embedded in extensive solidarity and communication networks, in mutual aid associations, markets, taxi stations and what has been described as 'bush telegraph', that provided them with an edge over their male counterparts and, during the fight against colonial occupation, turned them into major actors. While digital technologies change the nature of the territory, long-standing feminist activist practices continue to shape digital campaigns such as Guinéenne Du 21e Siècle.

This was powerfully highlighted to me when I interviewed Diérétou Diallo, the member of the Gdu21es collective behind the campaign. Recounting one of the most memorable moments from the campaign, she explained how for her one of the most meaningful tweets generated from the campaign was one that included a photograph of two young women in Siguiri, a rural part of northeastern Guinea, holding a handwritten sign that read '#guineennedu21esiecle 8 Mars 2016 Vive Les Femmes!' [#21stcenturyguineanwoman 8 March 2016 Long Live Women!]. While the post shared on Twitter (now X) on 8 March 2016 didn't include any other messages or commentary in response to the hashtag and campaign, it was particularly significant to her because it featured two women who did not have access to the internet directly themselves but who had heard about the campaign through word of mouth on their way to washing clothes at the local stream and wanted to participate. With the help of a third person, they had posted this photograph in support of Guinéenne du 21e Siècle, thereby exemplifying the mobilization power of feminist activist practices in what could be described as a continuation of what older generations described as the 'bush telegraph' in the digital era.

Much of the commentary within both local and international press regarding the Guinéenne Du 21e Siècle campaign emphasized the newness of the campaign's digitally focused methods, which they repeatedly framed as 'modern', part of the 'digital revolution' (Sagno, 2016), or reflected the collective's engagement for a 'new Guinea' armed with 'creativity and innovation' (*El País*, 23 April 2016). The Spanish newspaper, in the except shared above, further emphasized the campaign's non-disruptive approach that avoided sentimentalizing the issue. Framing it in performative terms as new and innovative, such understandings of the campaign strip it of its gender politics and the more radical potential reflective of women's strategic interests (Molyneux, 1985) it in fact contains. Part of this framing as a rupture from past political struggles stems from the fact that in Guinea, as in much of Francophone West Africa, the 'dreams of independence' that motivated many early independence movements are now often understood as a failure. Yet there is in fact much that this initial dream did achieve and that continues to live on (Kisukidi, 2017). A closer look at the user-generated content produced as part of #guineennedu21esiecle presents striking lines of continuation with the vast amount of literature on women's liberation produced during the First Republic, most notably by the Patrice Lumumba Press, the Guinean state's official press at the time. Publications such as *La Place De La Femme Dans La Société* (1977), presenting Sékou Touré's thoughts on women's roles in Guinean society under the revolution, present eerie parallels to the responses shared as part of the Guinéenne Du 21e Siècle campaign. This includes a set of photographs depicting women at work, at the pool, studying or taking part in research or military operations, with captions such as

> *Libérée des "tabous", la femme de Guinée est entrée au cœur des activités socio-culturelles qui lui assurent plus d'équilibre physique et mental.* [Freed from "taboos," the Guinean woman has entered the heart of socio-cultural activities ensuring her mental and physical health.]

> *La Guinéenne dans la recherche scientifique pour la protection et l'amélioration de la santé du people* [The Guinean woman taking part in scientific research for the protection and development of the people's health]

Mon métier: mon premier mari, celui qui ne me trahira jamais. [My work: my first husband, the one who will never betray me.]

These, alongside images of women at work, typing in positions almost identical to many of the photographs shared as part of the digital campaigns, highlight the historical continuities and the need to account for these historical continuities, as proposed by Jack and Avle's framework of feminist geopolitics of technologies.

Conclusion

Dominant narratives about digital feminist citizenship campaigns in Africa, and the Global South more generally, tend to emphasize the radical potential of digital technologies for advancing feminist causes. Such narratives, widely circulated in international media and often cloaked in celebratory language, emphasize the newness of digital activist practices. In Guinea, for instance, news coverage in both local and international media of recent digital feminist campaigns such as Gdu21es at the heart of this chapter have noted the new affordances provided by digital technologies to promote feminist causes. Moreover, such coverage has also tended to emphasize the potential of digital tools to herald or spearhead marginalized Guinean women into digital modernity, a techno-centric discourse often closely reminiscent of prior discourses of development leapfrogging that dominated public discussions regarding Africa in the 1990s and early 2000s (Bruijin, Nyanmjoh and Brinkman, 2009).

Drawing on digital and grounded ethnographic research I conducted in Guinea and online between 2014 and 2017, this chapter has sought to counter reductionist and techno-centric understandings of feminist digital citizenship campaigns by historicizing recent campaigns such as Gdu21es. Looking historically at the long line of feminist citizenship activism in Guinea beginning with the fight against French colonial occupation, it is clear that the campaign drew on long-standing feminist activist practices. These included highly efficient and nimble communication networks, mobilization power within non-elite and rural populations, and storytelling and slogan-crafting knowledge passed down generationally through women's associations and support networks. The mobilization of these long-standing activist practices, which have secured women's place at the forefront of political life in Guinea, was clearly at play in the digital campaign that coalesced around the hashtag #guineennedu21esiecle, even when the campaign itself was framed as a departure from older generations of feminist actors in Guinea.

As feminist scholars of digital technologies Jack and Avle (2021) remind us, such historicizing is important to move away from techno-determinist and fetishist approaches to digital technologies. It also firmly embeds these recent practices away from neoliberal celebrations of feminist digital practices as 'not too disruptive' and tame. Instead, such an approach highlights their radical potential, pluralizes feminist digital citizenship practices and disrupts the continued erasure and silencing of feminist citizenship activism, making firmly visible the 'body, race, gender' of digital campaigning (Kisukidi, 2017).

Note

1   https://dieretoudiallo.com/guineenne-du-21e-siecle/

Bibliography

Ammann, C. (2020), *Women, Agency, and the State in Guinea: Silent Politics (Edition 1)*, London and New York: Taylor & Francis, https://doi.org/10.4324/9780429199547.

Bentouhami, H. (2017) 'Notes Pour un Féminisme Marron. Du Corps-Doublure au Corps Propre', in A. Mbembe, and F. Sarr (eds), *Politique des Temps: Imaginer les Devenirs Africains*, Paris: Philippe Rey, 2019.

Beoku-Betts, J. and Adomako Ampofo, A. (2021), 'Positioning Feminist Voices in the Global South', in A. Adomako Ampofo, and J. Beoku-Betts (eds), *Producing Inclusive Feminist Knowledge: Positionalities and Discourses in the Global South (Advances in Gender Research, Vol. 31)*, Leeds: Emerald Publishing Limited, pp. 1–19.

Bergère, C. (2019) 'From Street Corners to Social Media: The Changing Location of Youth Citizenship in Guinea', *African Studies Review*, 63 (1):124–45.

Bruijn, Mirjam de, Nyamnjoh F. and Brinkman I., eds (2009), *Mobile Phones: The New Talking Drums of Everyday Africa*, Bamenda and Buea, CM: Langaa RPCIG.

Diabaté, N. (2020), *Naked Agency: Genital Cursing and Biopolitics in Africa*, Durham, NC: Duke University Press.

Diaw, A. (2009), 'La Femme Entre Ordre et Désordre Publics Les Ambiguités de la Modernité', *Diogène*, 228 (4): 50–9. https://doi.org/10.3917/dio.228.0050.

Goerg, O. (2007), *Perspectives Historiques sur le Genre en Afrique*, Paris: L'Harmattan.

Habermas, J. (1991), *The Structural Transformation of the Public Sphere: An Inquiry into a Category of Bourgeois Society*, Cambridge, MA: The MIT Press.

Irigaray, Luce (2002), 'La Question de l'Autre', *Labrys, Etudes Féministes*, 1–2. http://vsites.unb.br/ih/his/gefem/labrys1_2/irigaray2.html.

Jack, M. and Avle, S. (2021), 'A Feminist Geopolitics of Technology', *Global Perspectives*, 2 (1). https://doi.org/10.1525/gp.2021.24398.

Jolaosho, O.(2019), 'Singing Politics: Freedom Songs and Collective Protest in Post Apartheid South Africa.' *African Studies Review*, 62 (2): 6–29, https://doi.org/10.1017/asr.2018.16.

Jolaosho, O. (2018), 'Awkward Activisms: Gender and Embodied Mobilization in a Postapartheid South African Social Movement.' *Signs: Journal of Women in Culture and Society*, 43 (2): 425–48, https://doi.org/10.1086/693555.

Kisukidi, N. Y. (2017), 'Laetitia Africana: Philosophie, Décolonisation et Mélancolie', in Mbembe, A. and Sarr, F., *Écrire l'Afrique-Monde*, Paris: Philippe Rey, 51–69.

Kisukidi, N. Y. and Ribeiro, D. (2021), *Dialogue Transatlantique: Perspectives de la Pensée Féministe Noire et des Diasporas Africaines* (P. Anacaona, Trans.), Paris: Anacaona Editions.

Madenga, F. (2021), 'From Transparency to Opacity: Storytelling in Zimbabwe under State Surveillance and the Internet Shutdown', *Information, Communication & Society*, 24 (3): 400–421. https://doi.org/10.1080/1369118X.2020.1836248.

Molyneux, M. (1985), 'Mobilization without Emancipation? Women's Interests, the State, and Revolution in Nicaragua', *Feminist Studies*, 11 (2): 227–54. https://doi.org/10.2307/3177922.

Nyamnjoh, F. B. (2022), 'Citizenship, Incompleteness and Mobility', *Citizenship Studies*, 26 (4 5): 592–8. https://doi.org/10.1080/13621025.2022.2091243.

Pauthier, C. (2007) 'Tous Derrière, les Femmes Devant! Femmes, Représentations Sociales et Mobilisation Politique en Guinée (1945 – 2006)' in O. Goerg (2007) *Perspectives Historiques Sur Le Genre En Afrique*, Paris: Editions L'Harmattan.

Rajabi, S. (2021), *All My Friends Live in My Computer: Trauma, Tactical Media, and Meaning*, New Brunswick, NJ: Rutgers University Press.

Rivière, C. (1971), *Mutations Sociales en Guinée*, Paris: M. Rivière et Cie.

Roberts, T. and Bosch, T. (eds) (2023), *Digital Citizenship in Africa: Technologies of Agency and Repression*, London: Zed Books.

Touré, A. S. (1997), *La Femme dans la Société*, Conakry: Imprimerie Nationale 'Patrice Lumumba'.

Towns, A. R. (2022), *On Black Media Philosophy*, Oakland, CA: University of California Press.

Sagno, G. (2016), La Guinéenne Moderne, *BBC Afrique*, 10 March.

Schmidt, E. (2005), *Mobilizing the Masses: Gender, Ethnicity, and Class in the Nationalist Movement in Guinea, 1939-1958*, Portsmouth, NH: Heinemann.

Willems, W. (2023), 'The Reproduction of Canonical Silences: Re-reading Habermas in the Context of Slavery and the Slave Trade', *Communication, Culture & Critique*, 16 (1): 17–24. https://doi.org/10.1093/ccc/tcac047.

Chapter 7

TRANSFORMATIVE MOMENTS IN FEMINIST DIGITAL CITIZENSHIP IN POST-REVOLUTION EGYPT

Manal Hassan

Egyptian youth have innovatively utilized information and communication technologies (ICTs) for mobilization for decades. However, the January 2011 revolution sparked a tremendous explosion in online spaces, and women have contributed to the foundation of a significant share of these spaces (Alsherif, 2023). Furthermore, the surge in online spaces in the following years was proportional to the increased closing down of public spaces and intensifying crackdown on civil society following the military coup in 2013 (Mansour, 2017). Women, in particular, managed to carve alternative spaces in which they could construct a discourse that unites them. Focusing on issues in the private realm, they were able to contest social conventions without getting entangled in the polarized political landscape and risking being seen as a threat to the government (Elsheikh and Lilleker, 2021). In creating communities in which women are allowed to articulate their identities and needs, they started shaping a new discourse that defies the patriarchal system, which became the foundation of digital feminist citizenship in Egypt.

Despite the existence of some studies analysing specific campaigns or stories, or exploring how a particular digital platform was used as a case study of digital citizenship, if each of these stories is inspected on its own, the impact might not appear very significant. Moreover, if we only value the contribution of the feminist movement after the revolution on the basis of political representation and new legalisation, we might not be very impressed. Thus, this chapter focuses on collating the simple acts of everyday feminist digital citizenship in the many stories observed and documented by other scholars to examine them next to each other against the backdrop of the increasingly suffocating national struggles during that period. It examines how the affordances of offline and online spaces created since 2011 have laid the foundation for feminist digital citizenship in Egypt, looking at space affordances to Egyptian women at three successive periods, namely, the pre-revolution years, the revolutionary moment and after the military coup. Through the lens of affect theory, it traces the small shifts that have accumulated through the years, creating the foundation for feminist digital citizenship and contributing to strengthening the feminist movement in Egypt altogether.

Background and context

Digital context

DataReportal's country report on Egypt released in 2024 recorded its total population at 113.6 million, with a gender ratio of 49.4 per cent to 50.6 per cent (F:M). This report recorded 82.01 million internet users in Egypt, meaning an internet penetration of 72.2 per cent. Moreover, mobile penetration reached 97.3 per cent, and social media reached 40 per cent of the population, with a gender ratio of 38.6 per cent to 61.4 per cent (F:M) of the total social media users (Kemp, 2024).

Amid the growing economic crisis over recent years, numerous low-income families have sought alternative income sources. With high penetrations of mobile, internet and social media, many Egyptians have begun to explore generating income through advertising on video blogs and social media channels (Guergues, 2020). The popularity of video streaming platforms in Egypt was acknowledged by YouTube and sought after by TikTok – a newer player in the video streaming arena – which swiftly entered this lucrative market by establishing connections with state institutions and expanding its user base, which reached 7.2 million in 2019 (AFTE, 2021b). Unfortunately, video content by female vloggers is often frowned upon and reported for violating 'family values', 'societal principles' or 'public decency'.[1]

Legislative context

Egypt's record on freedom of expression has been in decline in recent years, ranking 170 out of 180 countries in the Reporter Without Borders World Press Freedom Index (RSF, 2024) and indexed as 'Not Free' in the Freedom House report (Freedom House, 2024). In December 2016, the Egyptian parliament passed a law to create three regulatory bodies for media: the National Press Authority, the National Media Authority and the Supreme Council for Media Regulation (SCMR), appointed by the president (Article 19, 2019; MOM, 2019), which are considered surveillance rather than regulatory bodies (AFTE, 2021b, 2022).

In August 2018, the Egyptian parliament passed two critical laws that eventually increased the state's control of online, print and broadcast media, namely, law no. 175 on combating Information Technology Crimes and law no. 180 on regulating the press, media and the SCMR; both laws increased the state's power to criminalize content creators on profoundly vague and abstract concepts of family values, principles of society and the security of the nation (CIHRS, 2020; AFTE, no date). Moreover, the new media regulation law considered any social media account, blog or website with more than five thousand followers a media outlet, thus regulated by the SCMR (BBC, 2018). Furthermore, in 2019, the Public Prosecution Office (PPO) established the Communication and Guidance Department, which includes the Monitoring and Analysis Unit, tasked with monitoring social media platforms and the Egyptian cyberspace altogether (ibid.).

Literature review

Despite a wealth of academic literature on social movements in Egypt before the January revolution in 2011 (Oweidat et al., 2008; Shehata, 2012; Hassabo, 2017, 2019; Mansour, 2017; Hassan, 2021), the literature on digital citizenship and the emergence of the Egyptian blogosphere has mainly focused on the political aspect of digital citizenship (Abdulla, 2005; Malky, 2007; Radsch, 2008; Al-Ani et al., 2012; Tufekci and Wilson, 2012). And while some scholars have explored the emerging voices of young Egyptian women (Otterman, 2007; York, 2009; Elsadda, 2010; Harp, Loke and Bachmann, 2014; EL Zein, 2015), only a few writers have captured the multilayered richness of the Egyptian blogosphere, recording the early experimentations and the initial encounters (Naji, 2010).

After the revolution, academic literature recorded how Egyptian women were pushing several boundaries, with scholars like Sorbera (2014) and Kamal (2016) observing how a new feminist citizenship was born, commencing the fourth wave of feminist mobilization in Egypt (ibid.).[2] On recounting the various milestones of feminist mobilization in the initial post-revolution years, scholars agree that the most significant accomplishments revolved around the consolidated efforts put into shaping the constitution and the initiatives that sprung up to counter systemic sexual violence (Kamal, 2016; Abdel Hadi, 2018), ultimately resulting in a 2014 amendment to the Egyptian penal code criminalizing sexual harassment (Bayat, 2021b; Alsherif, 2023).

While some literature has focused on documenting specific feminist initiatives, such as the establishment of the Coalition of Feminist Organisations in 2011 to push for women's political participation in drawing the post-revolution roadmap (Kamal, 2015), feminists' involvement in drafting the constitution in 2014 (Elsadda, 2015) and the launching of OpAntiSH to counter mass sexual assault (El-Rifae, 2022), other scholars have shifted their focus to the thriving of online spaces triggered by the revolution (Alsherif, 2023), the development of online micro public spheres in which women were empowered to confront societal norms, and the emergence of a parallel, feminist digital citizenship (Elsheikh and Lilleker, 2021). Scholars have observed how the fourth wave of feminist mobilization in Egypt revolved around the issues of women's bodies and rights (Kamal, 2016), noting how women's bodies became a political battleground (Letsinger, 2012; Abouelnaga, 2015; El Said, Meari and Pratt, 2015) and highlighting how body disciplining was used as a strategy to deter women from participating in protests (Hafez, 2014a; Abouelnaga, 2015). However, Abouelnaga (2015: 49) has argued that '[T]he more the female body was abused, the stronger and more solid the activism that was practised'.

While some scholars have observed a rise in feminist digital journalism celebrating women's achievements among various sectors and critiquing gender discrimination in the public sphere (Kamal, 2021), in written as well as audio format (Fox and Ebada, 2022), others have focused on the online gender backlash, arguing that the Egyptian manosphere[3] became more organized under Sisi's rule (Alsherif, 2023), with attacks ranging from aggressive comments on a post to threats of violence and rape to the content creator (Samir, 2020) or her loved ones (El Asmar, 2020). Unfortunately, there is very little academic literature that deeply analyses prominent feminist digital campaigns and hashtags in Egypt. Furthermore, there is a gap in studying how feminist digital citizenship has evolved in Egypt over time.

Conceptual framework

According to Roberts and Bosch (2023), being a digital citizen means using digital tools, such as mobile phones, the internet and social media platforms, to engage with your communities and navigate your social, economic and political realities. Combining this definition with Kamal's (2016: 5) view, considering that 'any act of women's dissent, regardless of its demands, carries a feminist dimension' and should be regarded as an expression of the agency of these women, I define feminist digital citizenship as any act of women's civic engagement utilizing digital tools to navigate their social, economic and political realities, even if some forms of this engagement might not have explicitly feminist agendas.

In this chapter, I utilize secondary data resources to analyse various significant moments of feminist digital citizenship next to each other. By examining the affordances of offline and online spaces for feminist citizenship at three successive periods, namely, the pre-revolution years, the revolutionary moment and after the military coup, I aim to portray the landscape of feminist digital citizenship in Egypt. This approach allows me to delve into the multifaceted nature of feminist movements within the digital sphere, highlighting how online spaces have become crucial arenas for advocacy, solidarity and mobilization. These data resources span academic literature covering one or more of these significant moments of feminist digital citizenship, in addition to grey literature in the form of reports by civil society organizations and media coverage.

Affect theory

Affect theory foregrounds subjective experience, feelings and emotions as ways of knowing, focusing on exploring how experiences, emotions and personal perspectives interconnect through encounters involving oneself, others and various dimensions of the space in which these encounters occur. In 'highlighting these moments of encounter, theories of affect draw our attention to the spaces of possibility – where change occurs, where we react, and where we begin to respond by producing new ways of being' (Kouri-Towe, 2015: 31). This provides new perspectives on transformational politics, helping us understand how power operates by turning to the forces that disrupt it (Jakimow, 2022).

To explain how new ways of being are produced, Kabeer uses Bourdieu's concept of 'doxa', which refers to deeply ingrained traditions and beliefs that exist without being questioned or debated. Kabeer observes how the world of doxa remains intact as long as people's subjective perceptions align with the possibilities available to them, and how it crumbles as soon as people perceive alternative possibilities of being and doing, exposing the arbitrary nature of the existing social order (Bourdieu, 1977, cited in Kabeer, 1999). A new discourse that offers possibilities for other ways of being, prodding people to question the status quo, is a social transformation in itself (Kabeer, 1999). How people interact with that discourse, what alternatives it triggers in their imaginations, and the shifts that alter the ways of being are the 'mundane victories of social movements' (Kouri-Towe, 2015: 25).

Affect theory's strength relies on a textured approach to reading the various forms of agencies, away from the binary of win/loss commonly used to analyse social movements. It is more concerned with the process of transformation than with evaluating the outcome of that transformation (Kouri-Towe, 2015). Gregg and Seigworth (2010: 2) observe that '[A]ffect is born in in-between-ness and resides as accumulative beside-ness'; it is this accumulative beside-ness that creates a discourse transforming individual feelings into a powerful force that binds the collective, guiding their actions and shaping their political identity (Chamberlain, 2016). Collectives formed through shared emotion foster a feminism that can adapt and grow with affect at its core, 'forming a cohesive series of relations and connections' (ibid.: 461).

Affordances

The theoretical concept of affordances was developed in the field of ecological psychology (Zhou and Xu, 2021), in which Gibson saw affordances as the possibility of 'an interaction between the physical properties of an object and the actions of a social agent' (Gibson, 1979, cited in Pruchniewska, 2019). For instance, a chair affords the possibility of sitting, and a car affords the possibility of mobility. Greeno (1994) saw the fundamental relation of the concept of abilities to carry a certain action to that of affordances, wherein an affordance connects the qualities of aspects of the environment to an interaction by an agent with certain abilities and vice versa. Since then, the concept of affordances has been adapted and used in various scholarly fields, such as the field of ICTs, wherein the concept has been used to highlight the unique traits of different communication channels and to observe how their interactive features affect human communication, for example, how mobile devices allow for freedom of movement for their users (Zhou and Xu, 2021).

Affordances can be seen as the potential actions availed to a given user group by a specific technology (Markus and Silver, 2008, cited in Henningsson et al., 2021). The same technology can afford other actions to other user groups; for instance, the mobile devices that afford freedom of movement to their users also afford citizen surveillance to security agencies. However, an affordance does not imply the actualization of the potential action (Stoffregen, 2000, cited in Henningsson et al., 2021). Here we have to note that a given technology affords different actions at different points in time, in other words, different versions of the same technology. Thus, when we speak of the affordances of a social media platform such as Facebook, we have to differentiate which version we mean –

the one that was launched, or the one that introduced pages, or the one that introduced chats, etc.

Tufecki (2017) notes how digital tools are reshaping our perception of space and time, redefining the world's structure by linking people across continents and conserving events, words and images that might otherwise be lost over time, focusing on how these tools promote the formation of collective movements by 'by allowing similar-minded people to find and draw strength from each other' (Tufekci, 2014). Digital tools, and social media in particular, afford speed to the propagation of new discourses, catalyzing discourse normalization (Chamberlain, 2016) by amplifying affect and increasing its intensity (Gibbs, 2001, cited in McLean, 2020).

This chapter uses the affordances lens to explore how the openings and closings of offline and online spaces before, during and after the revolution afforded Egyptian women with opportunities to question deeply ingrained traditions and alternative possibilities of being and doing. Using the affect lens, it examines how the accumulative beside-ness of these new possibilities became the foundation for feminist digital citizenship in Egypt.

Feminist citizenship in Egypt

This section considers feminist citizenship in Egypt in three successive periods: the decade before the 2011 revolution, the revolutionary moment from 2011 until the military coup in 2013, and the period starting after the military coup to the present.

Pre revolution (2000–2010)

The first decade of the millennium marked the third and final decade of Mubarak's rule. It featured the break of state monopoly on print and broadcast media, where the emergence of satellite channels such as Al Jazeera changed how Egyptians engaged with regional events, such as the 2003 war on Iraq (Tufekci and Wilson, 2012). This period saw the rise of a broad pro-democracy protest movement, the coalitions of diverse political factions, strong youth and student activism and workers' strikes that eventually 'coalesced in a revolutionary momentum' in 2011 (Hassabo, 2019).

Third-wave feminists

Kamal (2016) groups Egyptian feminist citizenship from the late nineteenth century to the present into four phases, according to their demands. These four phases – or waves – evidence a continuation in the Egyptian feminist movement since its beginning: the first wave started at the end of the nineteenth century and lasted until the fifties; the second wave was between the fifties and the seventies; the third started in the eighties and lasted until the revolution in 2011; and finally

7. Feminist Citizenship in Post-Revolution Egypt

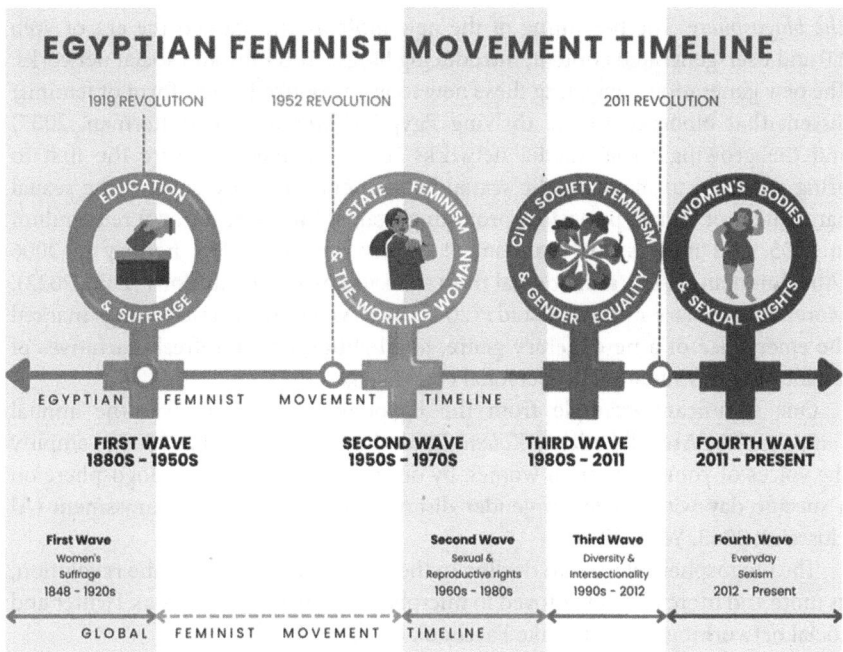

Figure 7.1 Timeline of the Egyptian feminist movement in relation to the timeline of the global feminist movement.

Source: (author), Data Source (Kamal, 2016), Icons: Canva Free Icons

the fourth wave started with the revolution, almost coinciding with the fourth wave of the global feminist movement (see Figure 7.1).

The feminist movement pre-revolution was mainly centred on the women's rights organizations established in the mid-eighties and during the nineties, which led several vocal campaigns around controversial women's rights issues, bringing them to the attention of the world in regional and international fora (Abdel Hadi, 2018). The Egyptian government, which had ratified the Convention on the Elimination of all forms of Discrimination Against Women in 1981 and hosted the International Conference on Population and Development in 1994, became threatened by the growing credibility of these organizations. In its attempt to minimize this threat, the state established the National Council for Women (NCW) in 2000, which competed with women's rights organizations for funding from international agencies (Tadros, 2014). The engagement of the feminist organizations with the transnational feminist movement was reflected in activism strategies during the first decade of the new millennium, which utilized international human rights platforms and regional networks and facilitated coalition building (Abdel Hadi, 2018). Furthermore, feminist dissent was visible in the various spaces of mobilization, whether in political activism such as in the pro-democracy protest movement or in workers' strikes (Abdel Hadi, 2018; Tadros, 2014).

The blogosphere The beginning of the new millennium marked the era of Web 2.0 and user-generated content, introducing blogs,[4] YouTube and social networks. The new generation navigating these new tools introduced a new form of feminist dissent that bloomed in the thriving Egyptian blogosphere (Otterman, 2007) and the growing social media networks. Egyptian bloggers were the first to bring attention to major public sexual harassment incidents, such as the sexual harassment of female protestors protesting against the constitutional referendum in 2005, and mass sexual harassment during the Eid al-Fitr holiday in 2006 (Abdelmonem, 2015), amid denial from state and media institutions (Zaki, 2022). Moreover, scholars such as ElSadda (2010) observed how the rise of blogs marked the emergence of a new literary genre, highlighting how the fresh narratives of female bloggers situated the 'personal center stage' (327).

One significant example from the Egyptian blogosphere was the annual campaign 'We Are All Laila', or *Kolena Laila*, that was launched in 2006 to amplify the voices of young Egyptian women by occupying the Egyptian blogosphere on a specific day with stories of gender discrimination and sexual harassment (Al Hussaini, 2008; York, 2010).

The blogosphere started to decline in the couple of years before the revolution, as more and more bloggers moved to microblogging platforms such as Twitter and social networking platforms like Facebook (Tufekci, 2017: 132).

The revolutionary moment (2011–2013)

After the revolution ousted Mubarak in 2011, the Security Council of Armed Forces (SCAF) assumed the executive role of governing Egypt during the transitional period. Despite promoting a narrative of the army protecting the people and safeguarding the revolution, by the end of 2011, the SCAF had strengthened its control over Egypt, arresting numerous prominent activists on unfounded charges, and it was responsible for the deaths and injuries of hundreds in violent incidents throughout that year (Badran, 2014; Sorbera, 2014).

In the squares

While scholars have repeatedly concluded that the Egyptian revolution in 2011 was a 'failed' and 'defeated' uprising (Aziz, 2014; Bayat, 2021a; Del Panta, 2022; Naeff, 2023), it was a turning point for most Egyptians, particularly the younger generation, forever changing their consciousness. It was an irreversible moment in which a different world could be imagined, when no one thought an alternative way of being existed (Mossallam, 2011). Egyptian women, in particular, could see an alternative world in which they were equal peers to men, they were seen as just human, and their bodies were respected (Magdy, 2012).

After toppling Mubarak, the general political mobilization leaned towards deferring feminist debates until democracy was achieved, but that did not deter the feminist movement from campaigning for their demands. As early as

February 2011, feminist activists decided to establish the Coalition of Feminist Organisations (Kamal, 2015), which issued statements expressing their opinions on how they envisioned the transitional period, commenting on the electoral law, the restructuring of the NCW, the formation of the committees drafting the constitution and the issues they wanted to integrate in the new constitution. While both the representation in the constitution committees and the endorsement of feminist demands were minimal, the new constitution did incorporate several significant new clauses, particularly addressing discrimination (Article 53) and violence against women (Article 11) (Abdel Hadi, 2018).

While Hafez (2014b) argues that women were excluded from behind-the-scenes negotiation meetings with the SCAF that shaped the political roadmaps, and Abdel Hadi (2018) observes that they were initially marginalized in party politics, in the subsequent years women were elected to leadership positions within political parties and professional syndicates (Kamal, 2016).

That moment in which women's bodies were respected was just a fleeting moment. In less than a month of toppling Mubarak, women who went to Tahrir Square commemorating International Women's Day were sexually harassed.[5] On the following day, 9 March 2011, more went to protest, and many were arrested and subjected to virginity tests by army doctors (HRW, 2011). This infuriated feminist activists, who staunchly campaigned against such atrocities (Seikaly, 2013). Later in the same year, video footage recorded at the dispersal of the Cabinet Sit-in showed a veiled young woman being dragged by army soldiers across the street. Still images from the video showed the woman dragged on her back by both arms, her black *abaya* ripped open and her upper body completely bare except for a bright blue bra,[6] with a soldier's army boot directly above her torso milliseconds before it hit her. In the video footage, the soldiers hit her several times, and one of them stomps her unresisting body. The images of the Blue Bra Girl went viral on the internet, pouring gasoline on a burning fire (Abouelnaga, 2015).

In 2012, a new method for body-disciplining emerged through mass sexual assaults, deliberately orchestrated to suppress feminist citizenship and restrict protest spaces (McRobie, 2014). Initially, different political factions, including leftist groups within the revolutionary movement, denied these assaults. Subsequently, when they acknowledged them, it was done discreetly to prevent any damage to the reputation of the idealized Tahrir Square (Abdel Hadi, 2018; Alsherif, 2023). These atrocious assaults and the shameful reactions – or lack thereof – from male comrades spurred women to organize differently. They established rescue teams such as Operation Anti Sexual Harassment (OpAntiSH) that created a tight operation, incorporating a hotline, scouts, on-ground intervention teams, escort drivers to hospitals and safe houses, and a documentation team recording the attacks and exposing the new tactics of using blades to rape women. OpAntiSH teams were mixed gender, asserting that women can be part of the on-ground rescue teams, even if they risk getting assaulted themselves (El-Rifae, 2022). Together with other groups, they collected testimonies, documenting five hundred cases of gang rape in two years (Nazra For Feminist Studies and the Center For Egyptian Women's Legal Assistance, 2014, cited in Mecky, 2018).

Off the squares

As early as 2011, we saw the birth of various daring initiatives, such as the regional 'Women's Uprising in the Arab World', which examined the commonalities of women's struggles in the uprisings that were sweeping the Arab world, expanding solidarity between women and advocating for women's participation in shaping the new political terrain (Alsherif, 2023). Another example is 'Girls' Revolution', established in 2012, which opened debates on very controversial topics, such as virginity tests conducted by families, marital rape and abortion rights (ibid.).

Post-coup (2013–2024)

Following a military coup in 2013 that overthrew Mohamed Morsi of the Muslim Brotherhood, the first democratically elected president post-revolution, SCAF member and defence minister Abdel Fattah el-Sisi was elected president in June 2014. Ever since the military coup, the state has adopted a militant approach in promoting a singular discourse, suppressing independent media, clamping down on civil society and eradicating any dissent, marking the most repressive period in Egypt's contemporary history (Hassan, 2021; EIPR, 2023). The gains made by civil society in the two years following the revolution gradually eroded, accompanied by a series of restrictive laws, the politicization of the judiciary and the absolute immunity granted to police and security services (Mansour, 2017).

New discourse With the closure of offline public spaces and the crackdown on on-the-ground dissent following the military coup in 2013, the daring and innovative feminist activism against sexual harassment was transferred to digital spaces. Whereas people debated its existence before the revolution, after the revolution, innovative campaigns to raise awareness about it, disputing the common reasoning that attributed it to women's behaviour and attire, were gaining momentum. Later, we started seeing the creation of closed spaces where women could discuss their personal experiences, provide mutual aid and solidarity, and begin building counter-narratives. Women-only Facebook groups, such as 'Confessions of a Married Woman', in which women anonymously posted taboo marital problems to get feedback from other married women, and 'FemiHub', in which young women migrating from smaller cities and villages to the capital – a growing trend since the revolution – to live independently from their families supported each other in their common challenges that ranged from economic independence to risks of exploitation as newcomers to the big city (ibid.).

Similarly, we saw the small communities that sprouted up online celebrating Egyptian women's natural curly hair and how they quickly evolved into a movement in which women encourage each other to defy societal norms. Egyptian women face intense pressures, bullying and discrimination that can affect hiring them in 'respectable' establishments (Geninah, 2022) if they do not straighten their unruly hair and conform to colonial beauty standards.

In parallel to the above, cyberspace saw a shift in websites and social media accounts targeting women, expanding from household-related content, such as cooking recipes and cleaning hacks, to other issues, such as reproductive and sexual health, sexual pleasure and the mental and psychological burdens of motherhood (Alsherif, 2023).

Initiatives such as 'WikiGender' contributed to localizing feminist theories, concepts and terminology with their archiving and translation efforts, in addition to setting up an online feminist dictionary (Alsherif, 2023). Similarly, 'The Sex Talk' initiative focused on discussing and accumulating knowledge on sexuality in Arabic (ibid.).

Sexual harassment and family values The year 2017 saw a major change in language, tone and discourse in relation to sexual harassment and women's bodily autonomy. Even before the beginning of the global #MeToo campaign, Egyptian women were taking social networks by storm in a campaign against sexual harassment with an Arabic hashtag that meant 'first time I was harassed my age was' (Alsherif, 2023). The global #MeToo campaign itself was echoed and amplified in several local campaigns naming and shaming a physician, a lawyer, a newspaper editor, a church clerk, a famous TV broadcaster and a former presidential candidate (Eltahawy, 2018; Zaki, 2022; Alsherif, 2023). One of these significant campaigns was the incident referred to as the 'Email Girl', which involved a widely circulated email from a young human rights activist accusing two of her previous colleagues in a leading human rights organization of sexual harassment and rape, in separate incidents. One of the accused was a prominent human rights lawyer, and the other was a presidential candidate in the 2018 presidential elections (ibid.).

The incident of the Email Girl had a significant impact, causing a rift within the Egyptian human rights movement (Eltahawy, 2018), but more importantly, it opened a debate on consensual sexual relations, even out of marriage (Alsherif, 2023). Scholars observe how the discourse on sexual harassment has significantly shifted post-2017 (Zaki, 2022) from the conservative narrative rooted in patriarchal values and populist rhetoric about protecting women and victim-blaming to emphasizing women's rights to a safe environment and asserting that harassed women should not be blamed for their attire, social behaviour or sexual activities (Alsherif, 2023).

The apex of this local #MeToo campaign was in 2020, when social media accounts such as Assault Police and Speak Up and websites such as *Modawanet Hekayat* were launched with the purpose of collecting sexual assault testimonies that might lead to the prosecution of rapists and perpetrators of sexual harassment,[7] at the same time breaking the silence on sexual assault (Zaki, 2022). The Instagram account @AssaultPolice was created by a university student to collect testimonies from the sexual assault victims of fellow university student Ahmed Bassam Zaki, who had a reputation of sexually assaulting his colleagues, but none of his victims dared to come forward for fear of defamation (Abuzaid and Sultan, 2022). With the anonymized testimonies shared on Assault Police, more came, and they were heavily shared across various social media platforms, getting more attention (ibid.). The state's NCW engaged with the case, declaring their complete support for the victims and encouraging more women and girls to send their testimonies (NCW, 2020a), and even filing an official complaint to the PPO on behalf of the victims (NCW, 2020b), who immediately ordered Zaki's arrest. Zaki was sentenced to eight years imprisonment in 2021 (Ahram, 2021).

After the solidarity shown to victims of Zaki's sexual assaults, a victim of a gang rape that had taken place in 2016 approached Assault Police (Abuzaid and Sultan, 2022). The perpetrators, who came from families with pervasive influence within the Egyptian state, had drugged the woman, taken turns raping her, marked her body with their initials and recorded the entire incident without any attempt to hide their identities, shamelessly sharing the video extensively afterwards (El Ammar, 2020). The NCW and the PPO were not as supportive in this case (ibid.).

Furthermore, the PPO was leading the state's crackdown on young female vloggers on the TikTok platform (Samir, 2020) who were reported by some male vloggers for wearing revealing clothing, dancing and encouraging other girls to open accounts like them to become brand models (ibid.). In a first precedent, these social media influencers were convicted of violating Egyptian family values and were handed sentences that ranged from two to six years in prison (AFTE, 2021a). Moreover, another vlogger was arrested for violating public morals, for posting a video recounting how she was raped (Samir, 2020). Discussions about consent, victim-blaming, class-bias and the evidentiary burden in cases of sexual violence resonated throughout the Egyptian cyberspace (Abuzaid and Sultan, 2022).

Personal status law In February 2021, the personal status bill presented by the cabinet to the parliament was leaked on social media (Elsadda, 2023). The leaked bill was a setback to the existing law, which feminist activists had been lobbying to reform, and included appalling provisions, such as allowing a woman's male guardian to terminate a marriage contract if he did not approve (ibid.). The Women and Memory Forum launched an online campaign called *Al-Weilaya Haqqy*, which translates to 'Guardianship is my Right', calling on Egyptian women to post their personal stories of the hardships they face in relation to their legal rights over their bodies and their children. The horrific stories of women's ordeals flooded the campaign's hashtag, spanning stories of women who could not make medical decisions for themselves or their children, mothers who could not enrol or transfer their children into schools, mothers who were not able to manage bank accounts they opened for their children, and more (Mada Masr, 2021). The campaign was adopted by other feminist organizations and activists, who called on their constituencies and followers to use the same hashtag, encouraging more and more women to share their stories, resulting in the hashtag becoming the second most popular hashtag in Egypt on Twitter one week after the campaign was launched (Elsadda, 2023).

Analysis

The varying affordances of spaces across time

Affordances are the potential actions availed to a given user group by a specific tool. I will use this lens to explore the opportunities afforded to Egyptian women with the openings and closings of offline and online spaces, before, during and after the revolution.

Pre-revolution (2000–2010) If we look at the decade before the revolution, we will find that the public space then afforded to feminists and women rights organizations the ability to join the transnational feminist movement and build coalitions at national, regional and transnational levels. Moreover, the public space then allowed for political dissent, so we saw more women joining various groups within the pro-democratic movement. And when we look at cyberspace then, we find that the rise of blogging allowed new and fresh female narratives to surface. Blogs afforded zero cost to set up and maintain a personal website, albeit requiring some technical literacy and familiarity with the English language, which meant that bloggers were disproportionately middle class.

The new medium gave bloggers the ability to experiment in topics, language and form. Moreover, it afforded to women the ability to blog anonymously under pseudo names and avoid pressures from their families and social circles.

The rise of Egyptian bloggers at the time of the pro-democracy protest movement gave them a spotlight since many of these bloggers were active in this political movement and quickly became citizen journalists covering what was happening on the ground, saying things as they were without the censorship of a news editor. This spotlight was initially acquired when bloggers were the only source of information and coverage of political demonstrations against the constitutional referendum and the sexual harassment of female protestors in 2005.[8] The fresh topics and the daring language of the blogs maintained traditional media's interest in featuring them.

The Egyptian blogs aggregator, which was a single website through which one could follow the updates of all Egyptian blogs, afforded a single portal from which to explore the whole of the Egyptian Blogosphere, discover new blogs and observe any new trends on this interesting new space. More importantly, the blogs aggregator facilitated the building of a close-knit community, which in turn promoted the emergence of a new discourse.

When the founders of *Kolena Laila* launched their campaign, the aggregator was flooded with blog posts of young Egyptian women telling different stories of gender discrimination they faced on a daily basis, and any follower of the blogosphere or the aggregator could not miss reading about gender discrimination in Egypt, even more so if you were a journalist covering the blogosphere. The same could be said of the incident of the mass sexual harassment in 2006; it could not be denied anymore.

The revolutionary moment (2011–2013) Now let us look at the affordances of Tahrir Square at the beginning of the 2011 revolution.[9] This space afforded to all who were part of it, particularly Egyptian women, new possibilities of being and doing. This is the space that enabled the world of doxa to crumble for many women. This space gave the feminist movement the opportunity to gain significant traction and resonate with a wider audience beyond the traditional circles of activists and academics, attracting diverse groups of women and allies from various social, economic, cultural and geographical backgrounds (Sorbera, 2014; Zaki, 2022).

The revolution afforded a spark of creativity and innovation in art, music, literature, self-organizing, political dissent and grassroots activism. Much of this completely changed after the military coup in 2013, with the crackdown on civil society and the almost absolute closure of all public avenues for street art, communal activities, political dissent and any sort of organizing. State violence against women, whether through the physical violence of virginity tests and the targeting of female protestors, or within state discourse depicting female protestors as loose, immoral and sexually promiscuous women, who spent days and nights camping in tents with men (Abouelnaga, 2015), caused sexual violence to escalate to 'epidemic levels' (McRobie, 2014).

As we move to cyberspace post-revolution, Tufekci notes the increasing usage of Facebook, arguing that it can be attributed to the way Egyptians equated having a Facebook membership with being connected to the networked public and getting exposed to 'the ideas and information circulated by political activists' (Tufekci, 2017: 133), even the SCAF communicated their decisions through their Facebook page first (ibid.).

Post-coup (2013–2024) Apart from the exposure to and circulation of political ideas, the Facebook platform afforded the creation of virtual groups with thousands of members such as 'FemiHub' and the curly hair community. Facebook groups allowed members to post anonymously, thus facilitating the discussion of thorny and taboo topics like in the 'Confessions of a Married Woman' group.

Hashtags became a recurring feature in most social networking platforms, affording ease and simplicity to new campaigns, to the point that some campaigns are now referred to by their hashtag.

Unfortunately, as state control increased in cyberspace, it gave the Egyptian manosphere the opportunity to thrive. The PPO began monitoring social media platforms and calling on honest citizens to report their suspicions (PPO, 2023), arresting female TikTok vloggers based on these complaints (PPO, 2020). Moreover, social networking spaces constrained several feminist initiatives with their content and community rules, as what happened with the banning of moderators of 'Women's Uprising in the Arab World' and the closure of 'The Sex Talk' page by Facebook for offensive or pornographic content (Alsherif, 2023). Such incidents afforded to young feminists the space to realize the patriarchal values ingrained in these digital platforms and connected them to digital rights activists.

7. Feminist Citizenship in Post-Revolution Egypt 141

Cohesion through affect

The affect lens helps us highlight the spaces of possibility where new ways of being are created, disrupting the status quo. When combined with understanding the affordances of the various spaces across time, the affect lens underscores how the accumulative beside-ness of these new possibilities became the foundation for feminist digital citizenship in Egypt.

As women started perceiving alternative possibilities of being and doing, they started exposing the arbitrary nature of the existing social order, opening debates on thorny issues. With the closure of public spaces after the military coup, online spaces and communities thrived. Egyptian Women began forming women-only groups, in which they contested social conventions, constructing a discourse that united them. 'Confessions of a Married Woman' and the curly hair groups are examples of communities that constitute closed safe spaces in which women are allowed to articulate their identities and needs, creating a new discourse that defies the patriarchal system. 'FemiHub' is another example of an online community built on feminist solidarity, creating a transformative space where participants transcended their ideological, religious and cultural differences (FRIDA, no date). These communities were shaping political narratives, contributing to their members' identity formation and their civic engagement, and learning democracy by practice and participation (Fig, 2011, cited in Elsheikh and Lilleker, 2021).

In cyberspace, activism against sexual harassment started shifting from addressing public harassment on the streets to confronting harassment in workplaces and within religious institutions. As more women began sharing their stories, it led to a collective awakening in which women started to recognize that the harassment they faced from trusted individuals was not normal and that they were not to blame for it. When the Email Girl sent her email, it was with the intention of cautioning other women about specific human rights defenders who were trusted within the activists' community (Eltahawy, 2018). She did not know how this email would trigger heated debates on taboo issues such as consensual sexual relations, within or outside marriage.

The accumulative efforts of the consecutive campaigns against sexual violence crossed several boundaries, unapologetically expanding debates on thorny terrains, such as extramarital sexual relationships and consent given under exploitation of power or authority, or intoxication, or coercion and manipulation, effectively turning activism against sexual violence into lobbying for sexual rights, a critique that was posed to third-wave feminists (Al Affifi, 2006). In the conservative Egyptian society that blames harassed women for their behaviour or attire, to debate a woman's consent despite being in 'the wrong' in the eyes of society for having sex out of marriage or consuming drugs or alcohol is a qualitatively different deed. This feminist discourse started to affect state discourse, as evident in Al-Azhar's statement that condemned sexual harassment irrespective of the woman's behaviour or attire (Abdel Hadi, 2018) and the public prosecutor's statement that accused a man of sexual misconduct without passing any moral judgements on the women who accused him, even though some of them stated that they sent him nude photos (Egypt Today, 2020).[10]

Communities and initiatives such as 'Confessions of a Married Woman' and 'The Sex Talk' and the various controversial campaigns against sexual assault, including marital rape, became the building blocks that reinforced and normalized a feminist discourse. However, discourse normalization would not have been attainable without the localization efforts of initiatives such as 'Wiki Gender' and 'The Sex Talk', among others, which in turn facilitated more local feminist reflections and writings.

The accomplishments of campaigns combating sexual violence contributed to more women shifting from blindly accepting the social order to questioning it and having a critical perspective. The momentum of these campaigns was easily channelled for the 'Guardianship is my Right' campaign, launched in refusal of the leaked disastrous personal status bill. The campaign's hashtag became a powerful platform, quickly filled with countless horrific testimonies by women from various backgrounds, revealing the widespread nature and 'the underlying arbitrariness of the given social order' (Bourdieu, 1977, cited in Kabeer, 1999: 441). As the hashtag gained traction, the overwhelming participation not only highlighted the pervasive injustice but also underscored the urgency for societal change, effectively rattling the patriarchal state. Within two weeks from the campaign's launch, the bill was forcefully pushed aside (Elsadda, 2023), paving the way for a broader societal debate involving multiple stakeholders. This halt in the legislative process opened the door for a deeper examination of the bill's implications and enhanced collaboration among diverse groups to shape a more informed and balanced approach moving forward.

Conclusion

This chapter set out to address the question of how offline and online spaces created since 2011 have laid the foundation for feminist digital citizenship in Egypt since the 2011 revolution. This was achieved by analysing the possibilities afforded by various spaces across time, before and during the revolutionary moment and after the military coup and the shrinkage of public spaces integral to it. Tracing the small shifts created by these possibilities beside each other, through the lens of affect, we could observe the evolution of feminist digital citizenship in Egypt.

While the revolutionary squares provided the feminist movement with fresh blood from various social, economic, cultural and geographical backgrounds, cyberspace allowed more women to become exposed to feminist and political ideas and to creatively experiment with self-organizing. Moreover, the new possibilities of being and doing facilitated by the revolutionary moment were enhanced in online spaces, which afforded women with safe spaces in which they could express their frustrations with the existing social order, and their solidarity with each other. This in turn, enabled women to problematize their personal perspectives into political mobilization, as in the case of the cautionary email that turned into activism for sexual rights, a major accomplishment of the fourth wave of the Egyptian feminist movement.

If we look at a specific story of feminist self-organizing or measure the participation in a specific campaign, their impact alone might not be significant. Moreover, the impact we observe might not be permanent or consistent, as in the case of official public statements on sexual violence, and might even get reversed, as observed in the increased gender backlash in Egypt, particularly in cyberspace. Through the lens of affect, we can see that the significance of these small acts of digital citizenship does not lie in their singular mundane impact but rather in their accumulative beside-ness, creating feminist discourses that propagate back and forth between the personal and the political, and becoming the foundation for feminist digital citizenship in Egypt.

At the beginning of the revolution, women activists refused to be labelled feminists. This has dramatically changed in the past decade; many individuals – women and men – and initiatives now adopt that label. To see the word 'feminist' casually used in mainstream media speaks volumes of the extent of the transformation. However, new affordances are often coupled with new constraints, and new forms of activism bring new challenges. Activism against sexual violence utilizing anonymous testimonies and public naming and shaming campaigns is encountering significant backlash, with attempts to discredit the survivors or the activists involved and sometimes taking them to court. We need to keep observing these small shifts as they unfold, for the startling insights lie not at the end of each story but in the richness of their continuity.

Notes

1 'Vloggers', or video bloggers, is a term used for content creators who post video content, including YouTubers and TikTokers.
2 As explained in the next section.
3 The term 'manosphere' refers to a wide variety of overlapping, mostly digital, communities of men who believe that men are naturally dominant but 'feel alienated in a changing world', thus they are against gender equality, and they promote a masculinist, misogynistic discourse.
4 Short for web-log, which initially started as personal websites with free-form diary-style text entries (blog posts), usually ordered in descending order from the most recent post to the oldest. The author of such a website was called a blogger.
5 *Tahrir* means 'liberation' in Arabic. Tahrir Square is one of the biggest and most central squares in downtown Cairo. It was a major focus and location for political dissent long before the January 2011 revolution. It was initially named Ismailia Square after Khedive Ismail, who founded the downtown district, but was later renamed Tahrir Square after the 1919 revolution.
6 Long all-encompassing dress.
7 *Modawanet Hekayat* is Arabic for 'Stories Blog'.

8   On the constitutional referendum day, the police also targeted journalists and photographers, confiscating their cameras.
9   Tahrir Square here is an idiom for all revolutionary squares in Egypt.
10 Al-Azhar is an Islamic institution and the highest religious authority for Muslims in Egypt.

Bibliography

Abdel Hadi, A. (2018), 'Assessing the State of the Women's Movement in Egypt 1980s–2018', in A Country Brief on Egypt, for the Regional Study on Assessing Women's Movement in Africa (1980s–2015), UHAI Investments LTD.

Abdelmonem, A. (2015), 'Reconceptualizing Sexual Harassment in Egypt: A Longitudinal Assessment of el-Taharrush el-Ginsy in Arabic Online Forums and Anti-Sexual Harassment Activism', *Kohl Journal*, 1(1). Available at: https://kohljournal.press/reconceptualizing-sexual-harassment-in-egypt.

Abdulla, R. A. (2005), 'Taking the E-train: The Development of the Internet in Egypt', *Global Media and Communication*, 1 (2): 149–65. Available at: https://doi.org/10.1177/1742766505054630.

Abouelnaga, S. (2015), 'Reconstructing Gender in Post-Revolution Egypt', in M. El. Said, L. Meari, and N. Pratt (eds), *Rethinking Gender in Revolutions and Resistance: Lessons from the Arab World*, London: Bloomsbury Academic & Professional.

Abuzaid, R. A. and Sultan, Y. (2022), 'On Social Networks, Anonymous Testimonies, and Other Tools of Feminist Activism against Sexual Violence in Egypt', *Journal of Middle East Women's Studies*, 18 (2): 301–10. Available at: https://doi.org/10.1215/15525864-9767968.

AFTE (2021a), *Short version - A Continued Isolation: The Annual Report on The State of Freedom of Expression in Egypt in 2020*, Association of Freedom of Thought and Expression. Available at: https://afteegypt.org/en/research-en/monitoring-reports-en/2021/03/02/21142-afteegypt.html.

AFTE (2021b), *The Role of App Owners, Association of Freedom of Thought and Expression*, Association for Freedom of Thought and Expression. Available at: https://afteegypt.org/en/the-role-of-companies.

AFTE (2022), 'Mass Surveillance: A Systematic Practice in State Institutions', Association for Freedom of Thought and Expression. Available at: https://afteegypt.org/en/highlight_en/2022/05/29/30634-afteegypt.html.

AFTE (no date) *Legislative context, Association of Freedom of Thought and Expression*, Association for Freedom of Thought and Expression. Available at: https://afteegypt.org/en/legislative-context.

Ahram (2021), 'Egyptian Court Jails Former Student for 8 Years Over Sexual Assault Charges - Politics - Egypt', *Ahram Online*, 11 April. Available at: https://english.ahram.org.eg/ NewsContent/1/64/408954/Egypt/Politics-/Egyptian-court-jails-former-student-for- years-ove.aspx.

Al Affifi, W. (2006), 'The Women's Movement, and Sexual Rights of Egyptian Women', *Tiba*.

Al Hussaini, A. (2008), 'Egypt: We are All Laila', *Global Voices*, 11 October. Available at: https://globalvoices.org/2008/10/11/egypt-we-are-all-laila/.

Al-Ani, B., Mark, G., Chung, J. and Jones, J. (2012), *The Egyptian Blogosphere: A Counter-Narrative of the Revolution*, ACM 2012 Conference on Computer Supported Cooperative Work26. Available at: https://doi.org/10.1145/2145204.2145213.

Alsherif, S. (2023), 'Egyptian Women Online: A Significant Imprint', Edited by M. Hassan, Translated by A. Abdel Hadi, *Motoon*, 28 July. Available at: https://motoon.org/resources/womenonline2011-2020-en.

Article 19 (2019), *Egypt: 2018 Law on the Organisation of Press, Media and the Supreme Council of Media, Article 19*. Available at: https://www.article19.org/resources/egypt-2018-law-on-the-organisation-of-press-media-and-the-supreme-council-of-media/.

Aziz, S. (2014), 'Bringing Down an Uprising: Egypt's Stillborn Revolution', *Connecticut Journal of International Law*, 27, 2014. Available at: https://doi.org/10.2139/ssrn.2480219.

Badran, M. (2014), 'Dis/playing Power and the Politics of Patriarchy in Revolutionary Egypt: The Creative Activism of Huda Lutfi', *Postcolonial Studies*, 17 (1): 47–62. Available at: https://doi.org/10.1080/13688790.2014.912188.

Bayat, A. (2021a), 'Introduction', in *Revolutionary Life*, Cambridge, MA: Harvard University Press (The Everyday of the Arab Spring), 1–6.

Bayat, A. (2021b), 'Mothers, Daughters, and the Gender Paradox', in *Revolutionary Life*, Cambridge, MA: Harvard University Press (The Everyday of the Arab Spring), 149–79.

BBC (2018), 'Egypt to Regulate Popular Social Media Users', *BBC News*, 17 July. Available at: https://www.bbc.com/news/world-middle-east-44858547.

Chamberlain, P. (2016), 'Affective Temporality: Towards a Fourth Wave', *Gender and Education*, 28 (3): 458–64. Available at: https://doi.org/10.1080/09540253.2016.1169249.

CIHRS (2020), 'Egypt: In Security Campaign to Protect Family Values, Public Prosecution Abdicates its Duty to Protect Citizens', Cairo Institute for Human Rights Studies (CIHRS). 24 June. Available at: https://cihrs.org/egypt-in-security-campaign-to-protect-family-values-public-prosecution-abdicates-its-duty-to-protect-citizens/?lang=en.

Del Panta, G. (2022), 'Reflections on the Failure of the Egyptian Revolution', *Middle East Critique*, 31 (1): 21–39. Available at: https://doi.org/10.1080/19436149.2022.2030982.

Egypt Today (2020), 'Egypt's Public Prosecution Releases Statement on Sexual Predator Ahmed Bassam Zaki', *EgyptToday*, 7 July. Available at: https://www.egypttoday.com/Article/1/89334/Egypt-s-Public-Prosecution-releases-statement-on-sexual-predator-Ahmed.

EIPR (2023) *Egypt's Worst Decade for Human Rights: Ten Years of a War on the Population Under the Guise of Fighting Terrorism | Egyptian Initiative for Personal Rights*, Egyptian Initiative for Personal Rights (EIPR). 24 July. Available at: https://eipr.org/en/press/2023/07/egypt%E2%80%99s-worst-decade-human-rights-ten-years-war-population-under-guise-fighting.

El Ammar, M. (2020), 'The "Fairmont" Case: Sexual Violence and Class Immunity', *Daraj*, 22 September. Available at: https://daraj.media/en/55598/.

El Asmar, F. (2020) *Claiming and Reclaiming the Digital World as a Public Space: Experiences and insights from feminists in the Middle East and North Africa*. Oxfam, 25 November. Available at: https://doi.org/10.21201/2020.6874.

El Said, M., Meari, L. and Pratt, N. (2015) 'Introduction', in M. El Said, L. Meari, and N. Pratt (eds) *Rethinking Gender in Revolutions and Resistance: Lessons from the Arab World*. London: Bloomsbury Academic & Professional.

EL Zein, R.A. (2015) *Between the blogosphere and the public sphere: Egyptian women bloggers before and after the January 25th revolution*. American University in Beirut.

El-Rifae, Y. (2022), *Radius*, London: Verso.

Elsadda, H. (2010), 'Arab Women Bloggers: The Emergence of Literary Counterpublics', *Brill*, 1 January. Available at: https://doi.org/10.1163/187398610X538678.

Elsadda, H. (2015), *Article 11: Feminists Negotiating Power in Egypt, Open Democracy*, 5 January. Available at: https://www.opendemocracy.net/en/5050/article-11-feminists-negotiating-power-in-egypt/.

Elsadda, H. (2023) 'Guardianship is my Right: Let us Tell our Story', *Tiba*, 20.

Elsheikh, D. and Lilleker, D. G. (2021), 'Egypt's Feminist Counterpublic: The Re-Invigoration of the Post-Revolution Public Sphere', *New Media & Society*, 23 (1): 22–38. Available at: https://doi.org/10.1177/1461444819890576.

Eltahawy, M. (2018), 'A #MeToo Moment for Egypt? Maybe', *The New York Times*, 13 March. Available at: https://www.nytimes.com/2018/03/13/opinion/egypt-metoo-email-girl.html.

Fox, K. and Ebada, Y. (2022), 'Egyptian Female Podcasters: Shaping Feminist Identities', *Learning, Media and Technology*, 47 (1): 53–64. Available at: https://doi.org/10.1080/17439884.2021.2020286.

Freedom House (2024), *Egypt: Freedom in the World 2024 Country Report*, Freedom House. Available at: https://freedomhouse.org/country/egypt/freedom-world/2024.

Geninah, F. (2022), 'Curly-Haired Women in Egypt are "Mocked in the Street", Says Entrepreneur Doaa Gawish', *The National*, 29 August. Available at: https://www.thenationalnews.com/lifestyle/wellbeing/2022/08/30/curly-haired-women-in-egypt-are-mocked-in-the-street-says-entrepreneur-doaa-gawish/.

Greeno, J. (1994), 'Gibson's Affordances', *Psychological Review*, 101, 336–42. Available at: https://doi.org/10.1037/0033-295X.101.2.336.

Gregg, M. and Seigworth, G. J. (eds) (2010), *The Affect Theory Reader*. Durham, NC: Duke University Press.

Guergues, A. (2020), 'YouTube, TikTok Help Low-Income Families Survive in Egypt', *Al-Monitor*, 13 August. Available at: https://www.al-monitor.com/originals/2020/08/egypt-youtube-tiktok-money-income.html.

Hafez, S. (2014a), 'Bodies That Protest: The Girl in the Blue Bra, Sexuality, and State Violence in Revolutionary Egypt', *Signs: Journal of Women in Culture and Society*, 40 (1): 20–8. Available at: https://doi.org/10.1086/676977.

Hafez, S. (2014b), 'The Revolution Shall Not Pass Through Women's Bodies: Egypt, Uprising and Gender Politics', *The Journal of North African Studies*, 19 (2): 185. Available at: https://doi.org/10.1080/13629387.2013.879710.

Harp, D., Loke, J. and Bachmann, I. (2014), 'Spaces For Feminist (Re)Articulations: The Blogosphere and the Sexual Attack on Journalist Lara Logan', *Feminist Media Studies*, 14 (1): 5–21. Available at: https://doi.org/10.1080/14680777.2012.740059.

Hassabo, C. (2017), 'Together, but Divided: Trajectories of a Generation of Egyptian Political Activists (From 2005 to the Revolution)', in M. M. Ayyash and R. Hadj-Moussa (eds), *Protests and Generations: Legacies and Emergences in the Middle East, North Africa and the Mediterranean*, Leiden: BRILL.

Hassabo, C. (2019), 'Coalitions for Change in Egypt: Bridging Ideological and Generational Divides in the Revolution', *Mediterranean Politics*, 24 (4): 491–511. Available at: https://doi.org/10.1080/13629395.2019.1639023.

Hassan, B. E. (2021), 'The Egyptian Human Rights Movement', in R. Springborg et al. (eds) *Routledge Handbook on Contemporary Egypt*, London and New York, NY: Routledge, 309–22.

Henningsson, S., Kettinger, W., Zhang, C. and Vaidyanathan, N. (2021), 'Transformative Rare Events: Leveraging Digital Affordance Actualisation', *European Journal of*

Information Systems, 30 (2): 137–56. Available at: https://doi.org/10.1080/0960085X.2020.1860656.

HRW (2011), 'Egypt: Military "Virginity Test" Investigation a Sham', *Human Rights Watch*, 9 November. Available at: https://www.hrw.org/news/2011/11/09/egypt-military-virginity-test-investigation-sham.

Jakimow, T. (2022), 'Understanding Power in Development Studies Through Emotion and Affect: Promising Lines of Enquiry', *Third World Quarterly*, 43 (3): 513–24. Available at: https://doi.org/10.1080/01436597.2022.2039065.

Kabeer, N. (1999), 'Resources, Agency, Achievements: Reflections on the Measurement of Women's Empowerment', *Development and Change*, 30 (3): 435–64. Available at: https://doi.org/10.1111/1467-7660.00125.

Kamal, H. (2015), 'Inserting Women's Rights in the Egyptian Constitution: Personal Reflections', *Journal for Cultural Research*, 19 (2): 150–61. Available at: https://doi.org/10.1080/14797585.2014.982919.

Kamal, H. (2016), 'A Century of Egyptian Women's Demands: The Four Waves of the Egyptian Feminist Movement', *Advances in Gender Research*, 21, 3–22. Available at: https://doi.org/10.1108/S1529-212620160000021002.

Kamal, H. (2021), 'Alternative Egyptian Feminist Journalism: The Case of Wlaha Wogoh Okhra', *Journal of the African Literature Association*, 15 (3): 413–28. Available at: https://doi.org/10.1080/21674736.2021.1935074.

Kemp, S. (2024), *Digital 2024: Egypt*, Data Reportal, 23 February. Available at: https://datareportal.com/reports/digital-2024-egypt.

Kouri-Towe, N. (2015), 'Textured Activism: Affect Theory and Transformational Politics in Transnational Queer Palestine-Solidarity Activism', *Atlantis: Critical Studies in Gender, Culture & Social Justice*, 37 (1): 23–34.

Letsinger, B. (2012), 'The Creation of a Revolutionary Icon', *Brandon Letsinger Writes*, 25 April. Available at: https://brandonletsingertravels.wordpress.com/2012/04/24/the-creation-of-a-revolutionary-icon/.

Mada Masr (2021), '#Guardianshipismyright: Women Call for Greater Legal Rights Over Their Children and Themselves', *Mada Masr*, 18 March. Available at: https://www.madamasr.com/en/2021/03/18/feature/politics/guardianshipismyright-women-call-for-greater-legal-rights-over-their-children-and-themselves/.

Magdy, Z. (2012), 'Egyptian Women: Performing in the Margin, Revolting in the Centre', *Open Democracy*, 23 January. Available at: https://www.opendemocracy.net/en/5050/egyptian-women-performing-in-margin-revolting-in-centre/.

Malky, R. A. (2007), 'Blogging for Reform: The Case of Egypt', *Arab Media & Society*, 5 March. Available at: https://www.arabmediasociety.com/blogging-for-reform-the-case-of-egypt/.

Mansour, K. (2017), 'Egypt's Human Rights Movement', *The Century Foundation*, 18 April. Available at: https://tcf.org/content/report/egypts-human-rights-movement/.

McLean, J. (2020), 'Feminist Digital Spaces', in J. McLean (ed.) *Changing Digital Geographies: Technologies, Environments and People*, Cham: Springer International Publishing, 177–201.

McRobie, H. (2014), 'The Common Factor: Sexual Violence and the Egyptian State, 2011–2014', *Open Democracy*, 6 October. Available at: https://www.opendemocracy.net/en/5050/common-factor-sexual-violence-and-egyptian-state-20112014/.

Mecky, M. (2018), 'State Policing: Moral Panics and Masculinity in Post-2011 Egypt', *Kohl Journal*, 4 (1). Available at: https://kohljournal.press/state-policing.

MOM (2019), Regulatory Authorities, *Media Ownership Monitor*. Available at: https://egypt.mom-rsf.org/en/context/law/regulatory-authorities/.

Mossallam, A. (2011), 'Why the Egyptian Revolution Matters To Us All', *Tinker Thoughts*, 10 November. Available at: https://tinker-thoughts.blogspot.com/2011/11/why-egyptian-revolution-matters-to-us.html.

Naeff, J. A. (2023), 'Remembering Defeat in Counter-Revolutionary Egypt', *Middle East Critique*, 32 (1): 53–69. Available at: https://doi.org/10.1080/19436149.2023.2168380.

Naji, A. (2010), *Blogs from Post to Tweet*. Arabic Network for Human Rights Information, p. 78. Available at: https://anhri.net/?p=7052.

NCW (2020a), 'The National Council for Women Announces its Full Support for the Girls Who are Victims of Harassment and Rape', *National Council for Women* Facebook Page, 3 July. Available at: https://www.facebook.com/ncwegyptpage/posts/ pfbid022m WNnM7p8AK3cy8ZuJAKHcqQ2ysapEv1JXw2aJTNkbgs4EBqz2Nppc9UJUwNT5Lul.

NCW (2020b), 'The National Council for Women Filed a Complaint to the Public Prosecutor Regarding the Famous "Social Media" Case', *National Council for Women* Facebook Page, 4 July. Available at: https://www.facebook.com/ncwegyptpage/posts/ pfbid02mTJQQ4bi6i5FsGGNvg7uPRPpJ4TdAXfp4XdU8ymW6wumerYkMQVvfqyn3 BqgHVUil.

Otterman, S. (2007), 'Publicizing the Private: Egyptian Women Bloggers Speak Out', *Arab Media & Society*, 5 March. Available at: https://www.arabmediasociety.com/ publicizing-the-private-egyptian-women-bloggers-speak-out.

Oweidat, N., Benard, C., Stahl, D., Kildani, W., O'Connell, E. and Grant, A. (2008), 'Kefaya's Successes', in *The Kefaya Movement*, Santa Monica, CA: RAND Corporation, 17–26.

PBN (2022), 'Curls Bounce Back in Cairo as Natural Hair Styles Become Trendy', *Premium Beauty News*, 17 April. Available at: https://www.premiumbeautynews.com/en/curls-bounce-back-in-cairo-as,20187.

PPO (2020), 'Statement from Public Prosecution in Case Number 4917 of Year 2020 El Sahel Misdemeanors', *Public Prosecution Office* Facebook Page, 23 April. Available at: https://www.facebook.com/ppo.gov. eg/posts/pfbid0LxUj1p35HMqe7zcJu9e8kGy768jGwGZMuhnjv BLBchZ9eLaBsHRTdhuhcRVzDmL5l.

PPO (2023), *Public Prosecution: Historical Documentation of the Most Prominent Work and Pleadings of the Public Prosecution 2019 -2023*, First. Public Prosecution Office. Available at: https://drive.google.com/file/d/1RUMbSyGfSaxKl-rTYeSOxhGjVbM5LfA7/view?usp=sharing&fbclid=IwAR2P6mmJwoCnFDsNLKao79 8BUIeF3IrYsxIvdmQfu5ceTtccrK9mpCLvjqg&usp=embed_facebook.

Pruchniewska, U. M. (2019), *Everyday Feminism in the Digital Era: Gender, the Fourth Wave, and Social Media Affordances*, Philadelphia, PA: Temple University Libraries.

Radsch, C. C. (2008), 'The Evolution of Egypt's Blogosphere', *Arab Media & Society*, 29 September.

Roberts, T. and Bosch, T. (2023), 'Introduction', in T. Roberts and T. Bosch (eds), *Digital Citizenship in Africa: Technologies of Agency and Repression*, London: Zed Books, 1–32.

RSF (2024), *Egypt, Reporters without Borders*. Available at: https://rsf.org/en/country/ egypt.

Samir, M. (2020), 'The Women of TikTok: Their Freedom is My Freedom', *Kohl Journal*, 5 (3). Available at: https://kohljournal.press/women-tiktok-their-freedom-my-freedom.

Seikaly, S. (2013), 'The Meaning of Revolution: On Samira Ibrahim', *Jadaliyya*, 28 January. Available at: https://www.jadaliyya.com/Details/27915/The-Meaning-of-Revolution-On-Samira-Ibrahim.

Shehata, D. (2012), 'Youth Movements and the 25 January Revolution', in B. Korany and R. El-Mahdi (eds), *Arab Spring in Egypt*. American University in Cairo Press, 105–24.

Sorbera, L. (2014), 'Challenges of Thinking Feminism and Revolution in Egypt Between 2011 and 2014', *Postcolonial Studies*, 17 (1): 63–75. Available at: https://doi.org/10.1080/13688790.2014.912193.

Tadros, M. (2014), 'Feminist Voices and the Regulation, Islamization and Quango-ization of Women's Activism in Mubarak's Egypt', *Voicing Demands: Feminist Activism in Transitional Contexts*, London: Zed Books.

Tufekci, Z. (2014), 'Capabilities of Movements and Affordances of Digital Media: Paradoxes of Empowerment', *Connected Learning Alliance*, 9 January. Available at: https://clalliance.org/blog/capabilities-of-movements-and-affordances-of-digital-media-paradoxes-of-empowerment/.

Tufekci, Z. (2017), *Twitter and Tear Gas: The Power and Fragility of Networked Protests*, New Haven, CT: Yale University Press.

Tufekci, Z. and Wilson, C. (2012), 'Social Media and the Decision to Participate in Political Protest: Observations From Tahrir Square', *Journal of Communication*, 62 (2).

York, J. C. (2009), 'Gender and Blogging in the Arab World', *Gender and Technology*, 8 April. Available at: https://archive.blogs.harvard.edu/genderandtech/2009/04/08/gender-and-blogging-in-the-arab-world/.

York, J. C. (2010), 'An Interview with the Founder of Kolena Laila', *Gender and Technology*, 7 January. Available at: https://archive.blogs.harvard.edu/genderandtech/2010/01/07/an-interview-with-the-founder-of-kolena-laila/.

Zaki, H. A. (2022), 'The New Feminist Movement Against Sexual Violence in Egypt 2011–2021', *Rowaq Arabi*, 27 (1): 43–61. Available at: https://doi.org/10.53833/GQIU6336.

Zhou, A. and Xu, S. (2021), 'Digital Public Relations Through the Lens of Affordances: A Conceptual Expansion of the Dialogic Principles', *Journal of Public Relations Research*, 33 (6): 445–63. Available at: https://doi.org/10.1080/1062726X.2022.2046585.

Chapter 8

DISMANTLING BOUNDARIES: MOZAMBIQUE'S TRAILBLAZING FEMINIST DIGITAL CITIZENSHIP

Lissungu Mazula and Dércio Tsandzana

In today's digital landscape, the transformation of relationships within the virtual sphere has enabled citizens to organize and support various activist actions. This holds significant importance in environments where women still face oppression, as civic groups advocating for women's rights have emerged. In Mozambique, despite the popular perception of feminism as an exported model, feminism is deeply ingrained in the country's history and socio-political terrain, reflecting the complex interplay of colonial legacies, liberation movements and post-independence nation-building processes. This chapter provides an overview of Mozambican digital feminism.

Furthermore, this chapter explores how urban women have been claiming their rights in the digital realm. It delves into the impact of digital platforms in empowering women, thereby fostering better digital citizenship and challenging traditional norms. There are many studies on gender and feminism in Mozambique, but few on digital feminism. For example, Arnfred (2011) reflected on feminist theorizing by rethinking gender (and sexuality) using the Mozambican case to reflect on gender politics, sexuality and matriliny. Arnfred (2011) also investigated different approaches to understanding gender and sexuality. Gender policies, from Portuguese colonialism to FRELIMO socialism and later neoliberal economic regimes, are thought to share certain fundamental assumptions about women, men and gender interactions.

This chapter is divided as follows: the next section presents the methodology; the third part discusses the historical context of being a woman in Mozambique; the fourth part presents the theoretical approach used in the chapter; section five is based on the digital landscape and citizenship in Mozambique; and the last part of the chapter is related to the case studies and analysis and discussion.

Context and country description

This chapter aims to reflect on how women have been claiming their rights in the digital realm in Mozambique and how the increased use of digital platforms has proved to be an ally in the resistance against forms of oppression and gender inequality. We will also analyse how the same digital platforms that have been amplifying the voices of many women in Mozambique and innovating the way in which resistance occurs can result in social exclusion.

In Mozambique, feminism emerged as a societal movement in the nineteenth century, aiming to confront the diverse types of oppression faced by women globally (Casimiro, 2004). Currently, there is an increasing proliferation of women-led associations in Mozambique. Casimiro (2004) highlights that many movements that arose following the establishment of the Organization of Mozambican Women (OMM) were formed with the purpose of pursuing opportunities for power and control or exploring alternative employment options, or were driven by the interest of international and funding agencies in establishing local partnerships in a more politically open environment.

Common factors that typically contribute to social isolation include social class, ethnicity, socio-economic status and cultural dynamics. Digital feminism, on the other hand, is still a topic underexplored in this context, particularly in how it can both challenge and reproduce these same structures of exclusion in online spaces. While this is true, digital feminism, even though not often discussed in academic circles in Mozambique, has started to hold significant promise for societal transformation. This potential arises from the increasing number of female-led initiatives that are utilizing digital platforms to challenge various forms of oppression and patriarchal expectations.

The emergence of digital platforms to promote collective feminism occurred in an environment where the intricate functioning of existing social movements became evident to others, along with their limited accessibility to diverse groups of women – social inclusivity. This phenomenon was evident in the way these organizations structured themselves and directed the spread of crucial messages on women's human rights to other subordinate associations.

Methodology

This chapter adopts a qualitative approach based on the use of case studies and profile interviews. We have analysed two cases studies: 'Sou Ntavase' [I am Ntavase] and 'Observatório das Mulheres' [Women's Observatory]. The first case is a virtual campaign that took place between 2020 and 2021. 'Sou Ntavase' was created against sexual violence and was carried out by the Socio-Cultural Association Horizonte Azul, in partnership with WLSA Mozambique and the Civil Society Forum for the Rights of the Child.

This campaign was inspired by a specific case, the sexual violence against a 10-year-old girl by a 36-year-old adult, whom we have given the pseudonym Ntavase to protect her identity. According to the organizers, 'Ntavase' is yet another child victim of a crime of sexual violence that continues to be practised in our society. Using digital platforms like Facebook and WhatsApp, the campaign aimed to inspire citizens to take a stand against these crimes and demand that the competent institutions act, becoming agents of change in order to curb this situation.

'Observatório das Mulheres' is a movement conceived because of a series of contemplations conducted by the women's collectives from September 2020 to March 2021. Although not a campaign as such, it is a national mechanism for representing a diversity of voices to empower women, ensure accountability and advocate for the removal of barriers that inhibit the realization of rights. It is an instrument for joint action and not a traditional association. The strategies centred on the priority concerning women's rights, aiming to comprehend their current position and the primary issues that existed, with a specific emphasis on the monitoring, evaluation and learning approach.

The two cases were chosen because they focus on the defence of girls' and women's rights through the virtual sphere. However, this chapter does not delve into digital netnography (Kozinets, 2019) but rather tries to understand the digital dynamics of the two case studies through targeted interviews. In fact, profile interviews are a qualitative research technique that entails conducting in-depth interviews with individuals to learn about their origins, experiences and viewpoints (Rubin and Rubin, 2012).

Profile interviews, as opposed to structured interviews, are semi-structured or unstructured, allowing for open-ended discussion and investigation of complicated themes. In this chapter, profile interviews allowed us to gather detailed information from Mozambican feminists, providing insights into their personal histories, motivations and daily experiences using digital platforms. The interviews were conducted with ten feminist activists that took part in the initiatives analysed in this chapter. All interviewees either hail from or have lived in the capital, Maputo, at some point in their lives.

All the respondents were chosen in snowball format through their direct or direct connection to the two case studies. We did not specifically ask the age of the interviewees, as this was not part of our methodological choice, but they all said they were young feminists, hypothetically in their mid-twenties and thirties. The interviews were conducted in Portuguese in April 2023, and all names were anonymized to protect the identity of those involved. We asked four specific and related questions:

What is your understanding of digital feminism? Is there really feminism in the virtual sphere?

How can the 'Sou Ntavase' campaign and the 'Women's Observatory' promote digital feminism in Mozambique?

How inclusive is the promotion of digital feminism in Mozambique?

What are the challenges of being a woman and using the digital sphere to promote rights?

Background: Female associativism in Mozambique

To understand women's resistance in Mozambique, it is important to understand the impact of their associativism in their participation in the national liberation struggle. Voluntary associations have experienced a substantial surge in number, starting from the 1980s and peaking in the mid-1990s in Mozambique. The observed phenomenon can be ascribed to a sequence of societal transformations that ensued as a consequence of the Western countries' adoption and imposition of neoliberal policies on peripheral nations. Frequently, these policies were implemented during democratic transitions and structural adjustment initiatives (Casimiro, 2014).

According to Cortesão and Casimiro (2021), Mozambique has experienced a culture threat aimed at feminists. One effective example of state repression of the feminist movement has been the portrayal of feminism as a foreign intrusion. The formation of these movements in postcolonial societies can be attributed to many circumstances. According to Tripp (2001), the rise of feminist movements in southern African countries can be attributed to a combination of external and internal reasons. These factors include the increased political freedom after colonial rule, the impact of women in conflicts, international pressure and the active mobilization of women themselves (Tripp, 2001).

However, in the context of Mozambique, as described by Casimiro (2001), this desire was expressed through the shared goal of improving living standards and ensuring the protection of women's rights at all levels (Casimiro, 2004). However, despite efforts to foster women's greater openness and engagement in political and civic spaces, feminism is still perceived as an inaccessible foreign model. The incident in March 2016, which culminated in the arrest of five feminist activists after attempting a public protest against the Ministry of Education and Human Development's policy that demanded that primary and secondary school girls wear full-length maxi skirts as part of their school uniform, exemplifies this notion. The emergence and expansion of the digital space has offered a safe space where women can raise their concerns about the unequal power relations and power distribution in Mozambican society. This space is, therefore, an opportunity to resist oppression. In other words, the campaigns analysed in this chapter illustrate how Mozambican women resist the constant criticism that they face when they demand rights. Likewise, this resistance is intended to show how the comparison of agency between movements based on a Global North dimension is the opposite of what happens in Mozambique. This reality can be divided into three stages.

FRELIMO as an ally: The origin of the Destacamento Feminino

During the 1960s and 1970s, the National Liberation Struggle (FRELIMO) demonstrated its commitment to promoting the rights and progress of Mozambican women, considering it an essential requirement for the development of the nation. The party contended that the freedom of women should coincide with the fight for independence from colonial control and the establishment of a new societal framework (Casimiro, 2001).

In this sense, women's activism and mobilization flourished during Mozambique's colonial area, as women found in FRELIMO favourable conditions for their integration and played a significant role in the fighting against Portuguese colonialism for national liberation. The involvement of women in the National

Liberation Struggle led to the creation of the Women's Detachment of the People's National Liberation Army in 1965. This movement was created at the request of women faced with the need to defend and mobilize populations in liberated areas, or in areas still controlled by colonialism (Casimiro, 2001).

Prior to their profile update, women were responsible for household chores related to the organization of production, specifically targeted at assisting military efforts and caring for children whose parents were engaged in fighting activities and unable to attend to them. With the introduction of the Destacamento Feminino, women began to actively participate in the struggle for freedom. Women with diverse backgrounds joined the FRELIMO through the Destacamento Feminino as warriors, organizers and propagandists. The liberation movement instilled a sense of unity and common purpose among female comrades (Bonate and Katto, 2021). Despite confronting prejudice and misogyny within their own ranks, Mozambican women emerged as critical agents of change in the fight against colonialism. While the Destacamento Feminino had a significant role in advancing the status of women as active individuals in the front line to end colonialism, it was insufficient to sustain their gains much longer due to the opposition to gender equality and to feminist leadership in public life.

Even though an exclusive women's division was established to offer military training to women and girls, their experiences still diverged from those of men, as sexual assault and other sorts of violence against women were prevalent within the army, and women frequently faced restrictions on their involvement in civilian affairs (Karberg, 2015). Subsequently, the OMM was established to respond to women's demand for equal participation in public life.

Organização da Mulher Moçambicana

Although created by members of the Destacamento Feminino, the OMM was controlled by a group of women with no history of national liberation who were in neighbouring countries. This implied a shift in the general perception of women's roles, as it was recognized that involving women in organized social production was crucial for challenging the traditional limitation of women to the domestic sphere, where they were subject to patriarchal authority. It was also acknowledged that gender equality and the liberation of women could only be achieved within this framework (Arnfred, 2011).

Through the OMM, women were encouraged to work for money in the first decade after independence, but they were mostly restricted to the subsistence economy, particularly in rural areas. The OMM upheld the party line, describing women as 'natural' caregivers. The achievement of independence in 1975 thus was a watershed moment in Mozambique's history, opening up new chances and difficulties for women's growth. However, the post-independence era was marked by economic insecurity, civil unrest and the legacy of colonial exploitation, which disproportionately impacted women.

Structural disparities remained, limiting women's access to education, healthcare and economic opportunities. Furthermore, the militarization of society during the civil war worsened GBV and insecurity among women. Many women were only able to take advantage of new opportunities during the 1990s' political and economic liberalization (Bonate and Katto, 2021) as the reconfiguration of the democratic system gave space for a political debate on women's rights (Osório, 2010) and as a result of the Beijing Conference held in 1995, which emphasized a women's empowerment agenda. For Mozambican women, this implied more political participation and representation. In fact, since the marginalization era established by the founding elections of 1994, much has been done by women's movements fighting for their independence and emancipation. For example, feminist movements in Mozambique have been instrumental in campaigning for gender equality, women's rights and social justice. These movements have emerged in response to several kinds of Mozambican women's oppression, which include patriarchy and economic marginalization.

The emergence of new (digital) feminist movements

In addition to the historical dimension of women's movements in Mozambique, there is a third wave represented by the democratic opening and the evolution of the media in the country. The first example is Fórum Mulher [Women's Forum], founded in 1993. It is a significant feminist organization dedicated to enhancing the rights and well-being of women in Mozambique. Fórum Mulher promotes women's autonomy and solidarity and advocates for women's economic, social, reproductive and political rights through coalitions at the local, regional and national levels.

The second prominent example is Observatório das Mulheres [Women's Observatory]. Founded in 2021, the Women's Observatory is a coalition of NGOs working to protect women from GBV; promote a broad, permanent and safe space for reflection, influence and monitoring of the situation of women in Mozambique; create a space to represent women in an equitable manner; and effectively allocate resources to respond to their diverse demands and realities. Using social media networks like Facebook, the Women's Observatory emerged as a national mechanism for representing a diversity of voices with the aim of empowering women and advocating for the removal of barriers that inhibit the realization of rights.

Understanding feminist digital citizenship

According to Roberts and Bosch (2023), digital citizenship uses mobile and internet technology for civic and political engagement. In Africa, these technologies allow quick content sharing and help citizens form organizations and share information without media or politics. These opportunities are, however, limited by uneven access to digital devices, connectivity and digital literacy. Despite these restrictions, millions of Africans use digital technology for civic and political engagement (Roberts and Bosch, 2023).

On the one hand, Oyedeme (2020) argues that the concept of digital citizenship is not new, as it has been applied before to draw attention to the importance of the Internet to citizens' ability to participate in society. Choi (2021), on the other hand, believes that the current approach to digital citizenship includes political activism and critical perspectives but lacks emphasis on critical elements, including marginalized groups, which are often excluded from traditional citizenship. This approach challenges existing digital citizenship by recognizing and addressing unequal power relations and social inequities on the internet and social media, promoting practices for social justice.

The internet has become a crucial platform for women to express themselves and voice their concerns. According to Mpofu (2016), online spaces allow individuals to establish and affirm their equal status in society and politics. These digital platforms empower marginalized individuals to share their stories and create 'cyber communities', where the internet fosters the development of personal opinions, alliances and social identities (Mpofu, 2016: 279). In essence, the internet has the potential to challenge traditional hierarchical power structures, providing women with a means to resist oppression.

Although gender is a crucial analytical category for understanding social relations and power imbalances, the literature on digital realms has been lacking, not only in the Anglo context but also in Lusophone contexts like Mozambique. According to Koutsogiannis and Adampa (2012), this gap exists because the broader literature often treats gender separately from other economic and sociocultural factors, overlooking contexts that may create contradictions in gendered practices. Furthermore, there is limited focus on how identity is constructed in computer-mediated communication for both women and men (Koutsogiannis and Adampa, 2012).

Theoretical approach

This chapter uses intersectionality theory (Crenshaw, 2004) to examine how various aspects of identity intersect and shape feminist citizenship in Mozambique. It explores how women activists continue to innovate and create solutions to address gaps in the feminist movement within the country. To achieve this, it is important to discuss key concepts such as cyberfeminism. As defined by Paasonen (2011), cyberfeminism refers to the feminist use and understanding of ICTs in both practical and theoretical contexts. According to Paasonen (2011), cyberfeminism is closely linked to the movement known as third-wave feminism, which is marked by its use of sarcasm and celebration of variety. However, cyberfeminism also distinguishes itself from traditional feminism, being considered as an alternative to it. Paasonen (2011) criticizes the cyberfeminist movement for its failure to effectively address power dynamics and inequality, despite its focus on diversity and humour, arguing that while women are encouraged to develop their own cyberfeminist ideas, there is often a lack of reflection on how power, location and difference operate within cyberfeminist networks.

In this research study, we will use this description, as we argue that a select group utilizing digital technology has significantly contributed to the emergence of an alternative form of resistance in Mozambique. Second, while the notion of patriarchy has consistently been a central theme in intellectual and feminist works (Eisenstein, 1979; Coward, 1983; Agarwala, 1988), there is also controversy regarding its precise definition. Aguiar (2000) argues that patriarchy can be seen as a power system similar to slavery. This perspective suggests that recognizing patriarchy as a problem leads to a set of normative demands aimed at addressing situations of unjust power within the family and society at large (Aguiar, 2000). However, according to Haraway (1988), patriarchy is a lengthy historical development that relies on an uneven allocation of power.

This system not only subjugates women but also marginalizes individuals based on many forms of oppression, including nationality, race, socio-economic class and religion. In contrast to Haraway, hooks (1981) argues that patriarchy refers to the authority that any male, regardless of his race or socio-economic status, exerts over a woman. hooks asserts that this power dynamic is inherently masculine, as it involves the use of force to demonstrate dominance over another individual. In hooks's (1981) analysis, white patriarchy is identified as exerting a far stronger influence on Black women. The author argues that white feminism has made no effort to address this issue.

For the purposes of this chapter, we use the latter concept because we realize that in Mozambique, the various women's social groups that have proliferated over time were first able to exist and create sustainability under the umbrella of patriarchy, i.e., the ruling party, so much so that we hypothesize that after the OMM and a few other initiatives, such as AMODEFA and Fórum Mulher, we will probably have a hard time talking about other women's social movements in the country. Another concept we aim to elucidate is the notion of resistance. Recently, there has been a notable surge in the scholarly investigation of resistance. However, there is a lack of consensus over its precise definition. For example, Lilja (2022) attributes this outcome to the deficiencies in the comprehensive understanding of the diverse and frequently interrelated forms of resistance within the domain of academic research. Freire (1983) defines resistance as the reaction that occurs when an individual belonging to a disadvantaged group becomes aware of their identity and recognizes their oppression. Subaltern resistance necessitates conscientization, which entails the intensification of consciousness.

Contemporary scholars perceive resistance and power as interconnected concepts, viewing them as two sides of the same coin. They argue that resistance should be approached as a holistic term that encompasses various manifestations, including everyday, serial and organized resistance, as well as the connections between them (Lilja, 2022). Foucault (1991) argues that various forms of power are associated with different forms of resistance and might potentially strengthen preexisting ties. Power and resistance are considered mutually exclusive elements (Lilja, 2022), suggesting that not only does resistance confront or incite power but also that, at times, engaging in resistance can reinforce the dominant power dynamic (Gaventa, 2003).

Digital citizenship in Mozambique

Digital citizenship has been one of the topics that is marking the debate in Sub-Saharan Africa (Robert and Bosch, 2023). In Mozambique, the discussion about the role of digital media in still limited (Joanguete and Tsandzana, 2023), even if young people have been using social media networks to do politics (Tsandzana, 2018). According to DataReportal (2025), Mozambique had 6.96 million internet users at the beginning of 2025, with an internet penetration rate of 19.8 per cent. Mozambique had 3.70 million social media users in January 2025, accounting for 10.5 per cent of the entire population, illustrated in Figure 8.1.

According to the same source (DataReportal, 2025), these user numbers show that 28.2 million Mozambicans did not use the internet at the start of 2025, implying that 80.2 per cent of the population was offline at the start of the year. In early 2025, Mozambique had 17.7 million active cellular mobile connections, which accounted for 50 per cent of the total population. According to data revealed in Meta's advertising resources, Facebook had 3.7 million users in Mozambique as of early 2025. Figures revealed in Meta's own tools show that Facebook's potential ad reach in Mozambique increased by 500,000 (15.6 per cent) between January 2024 and January 2025.

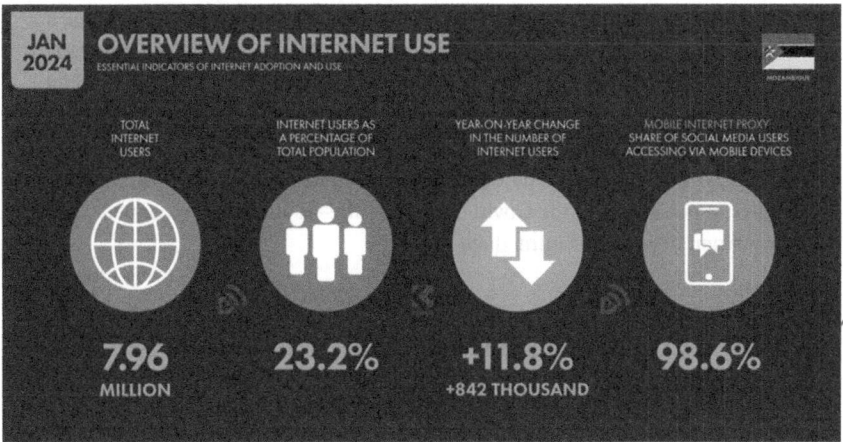

Figure 8.1 Internet use in Mozambique (free to use, DataReportal 2025)

In Mozambique's literature, for instance, the debate around digital citizenship has been limited to the focus on youth's political participation in elections. As argued by Tsandzana (2020), the digital realm has favoured a new alternative way of voting in which citizens, with particular focus on the country's youth, have been using it as an innovative means of exercising this participation, where they can therefore find greater visibility. Despite the increasing instances of technological violence, such as cyberbullying, online harassment and digital surveillance, in general, different forms of political action and empowerment are continuing to grow and develop, particularly for the feminists in Mozambique. Women are finding ways to exercise their political rights and engage in political activities even in the challenging environment created by technological violence.

Discussion: Feminism and the digital era in Mozambique

This section covers different perceptions of how to define digital feminism. In fact, using their own experience and practice, we asked the interviewees to explain how they understand digital feminism. Two of the interviewees noted that

> Feminism is a political and social movement that defends women's rights, seeking equality between men and women with social justice. In the digital context, women find their voice through access to these digital mechanisms. Although women are the majority in number, we still face inequalities and are victims of various types of gender violence, often rooted in religious and cultural norms that see us as objects for purchase and pleasure. In digital feminism, giving voice to these problems is fundamental, but it is also important to propose solutions.
> (DM, June 2024)

> For me, the approach to feminism in the digital sphere is complex because most feminists are uncomfortable with the digital space. They see the Internet as a limitation because feminism is not inherently digital; it's something you can't fully experience online. Many people advocate for women's rights in the digital sphere, but they don't really experience these problems firsthand. Moreover, the digital realm creates a fragmentation and division between physical and digital spaces. However, we can't deny that the digital space is important for women's struggles, especially after the Covid-19 pandemic – it allows us to connect with other women beyond our immediate sphere.
> (WS, June 2024)

Both interviews show that feminism can appear according to each woman's place of speech and perception and according to women's desires or the spaces they have to express their feelings – we cannot speak of a consensual perception. Equally, we note that there are still challenges of connectivity and lack of trust in the digital space. This is consistent with intersectionality theory, which emphasizes how various dimensions of identity – such as class, racism and access to technology – influence women's experiences and connections with feminism. In Mozambique, digital disparities exacerbate the situation, as connectivity barriers and distrust in online venues limit feminist expression.

Case study 1: #SouNtavase

> I think this campaign was extremely necessary because we women use the digital space a lot to raise awareness about women's rights. From there, we can empower our voices.
>
> (EN, June 2024)

Starting with the interview above, in this section we want to explore the level of engagement in some of the existent campaigns. Ntavase was the pseudonym assigned to a girl who survived sexual violence. As a consequence of the sexual assault, she had a profound lesion to the perineum, leading to a total rupture of the tissue connecting the anus and vagina. This injury resulted in substantial harm, rendering the expulsion of excrement through the anus impossible. Consequently, a surgical intervention known as colostomy was conducted to establish an alternate pathway for the removal of excrement through the abdominal wall.

The family needed help to cover the costs of food, diapers and other essential items to ease the suffering and the healing process. The case garnered significant attention from civil society and female activists, who utilized Facebook as a platform to advocate for justice on behalf of Ntavase. The offender has been sentenced to a twenty-four-year term of imprisonment, which is the maximum penalty permitted by the penal code in Mozambique. In addition, he was obligated to pay a sum of 150,000 meticais as financial restitution to the Ntavase family.

The campaign was widely disseminated on WhatsApp with the creation of specific groups and on Facebook via hashtags such as #SouNtavase [I am Ntavase], #fuiviolada [I was violated], #exijojustica [I want justice] and #exijorespeito [I want respect]. One of the interviews noted that #SouNtavase was a good campaign that expanded women's voices:

> Adopting the digital sphere was a good choice because we knew that in the traditional space there would be no response, and we felt that in the digital sphere we could be heard. The girl's mother tried to approach the television, but to no avail, so we decided to use Facebook. Our campaign started on WhatsApp, and from there we managed to mobilize support. We took her to the hospital so she could receive care for the after-effects of the sexual violation she suffered. We then managed to take collective action to bring the perpetrator to court. The campaign was primarily to protect the survivor's identity – it was a way of initiating a dialogue about sexual violence and protecting girls.
>
> (DA, June 2024)

The aforementioned interview reveals how a campaign that had a specific objective could have become a broad movement that encompassed a debate on sexual violence suffered by women in general. This insight highlights the multifaceted character of the feminist movement in Mozambique, as digital platforms open up new opportunities for justice and communal mobilization. While traditional media frequently operates within institutional and structural constraints, digital activism, founded in cyberfeminism, provides a more immediate and inclusive place for women to express their experiences. However, access to these digital tools is not consistent, as structural inequalities continue to influence how and how people participate.

The potential for a targeted campaign to grow into a larger movement demonstrates how dynamic digital feminist activism is, adjusting to the demands and lived reality of the women who participate in it. This flexibility challenges conventional narratives of female struggle, emphasizing the digital world as both an opportunity and a place of contestation. In another development, one of the interviewees said that the fact that it was digital increased the chances of having the rapist punished, which would not have been possible if the option had been traditional media, such as television.

> It was a very important campaign that we must consider, as it highlighted a problem by identifying and punishing the rapists of girls. The faces of the rapists were exposed, which allowed us to succeed with our campaign. It was timely because it forced public and justice institutions to take real action in defense of women's rights. The campaign wouldn't have had the same effect on the street, so the digital sphere was crucial for achieving success.
>
> (HM, June 2024)

Case study 2: Women's Observatory

Despite not being a largely digital campaign, the Observatório das Mulheres initiative emerged to counterbalance the traditional way in which women's organizations have emerged over time in Mozambique. However, one of the interviewees spoke about the limited reach of this (new) movement:

> I know that the Women's Observatory has run several campaigns in the digital space, but the problem is that some campaigns are sporadic and respond to specific issues. Ntavase is the result of a specific case, and from there, it has given a voice to others, creating more sustainability. Other campaigns are just events that spark their creation, but they don't have a long lifespan.
>
> (EN, June 2024)

The interview above reveals a debate that often takes place in the digital sphere, which has to do with the sustainability of some campaigns. The durability and impact of these campaigns is still a challenge that cannot be ignored. This difficulty alludes to broader contradictions within digital feminist advocacy, in which exposure does not always convert into long-term progress. From an intersectional standpoint, sustainability is dependent on access, involvement and the ability to link online activism with offline action. While digital spaces magnify voices, institutional constraints continue to decide which campaigns endure and which fade, emphasizing the importance of long-term impact initiatives.

Another interview addressed the issue of low humility in the digital sphere:

> I know there is also the Women's Observatory, but I don't really understand what they do. I'm not based in Maputo, and outside the capital, we have different agendas. Our feminism is not inclusive because many women are more privileged than others. Equally, many of these women should know how to give without expecting anything in return. I feel there should be humility in the way we approach feminism in the digital space, without bringing prejudices based on our centralized experiences. Women don't always need to be represented; they should also know how to represent themselves.
>
> (AA, June 2024)

In another development, we asked about the limitations women encountered in the digital space, as well as the dangers of using the Internet in everyday life. Many clearly stated that they did not feel protected in using the digital space.

> The big challenge for this type of campaign is the cost of the internet. We also have people in the digital space who still advocate violence and expose women without contemplation. While it is a space where we can do our activism in favour of women, it remains hostile to women. There is no effective way to protect women in the digital space. Additionally, we face attacks by individuals, including hackers, targeting women.
>
> (DA, June 2024)

> The biggest challenge is the retaliation that women suffer, especially digital violence against women, not to mention the harassment they still experience in their daily lives. Many women are attacked for promoting their rights in the digital sphere. These attacks are not only against transgender women but also lesbians. Women are forced to delete comments every day from men who attack their integrity. Another challenge is the cost of the internet, whether in urban or rural areas. Finally, women are not safe to carry out their activism.
>
> (WS, June 2024)

> Feminism in the digital sphere is still in its infancy; it is not as popular as it might appear. It is something that is gradually growing, but due to the nature of a patriarchal society, struggles in the digital sphere are often overlooked. Many people criticize based on their cultural and traditional beliefs. We see a lot of negative comments on social networks that do not support these kinds of campaigns. It is evident that the digital space often mirrors the physical one, where feminists also face significant criticism. There is a lack of understanding

of feminism's role in our society. The societal norms contribute to this situation, but there is a need for greater unity among us as women.

(HM, June 2024)

The three interviews quoted above reveal how women feel within the virtual sphere, which is still dominated by scenarios of harassment and patriarchal discrimination, but also situations that limit the reach of the digital sphere for women outside the connectivity circuit. It exemplifies the dual reality of digital spaces for women in Mozambique: while they provide new avenues for expression and activity, they are nevertheless governed by patriarchal power dynamics and technological marginalization. Intersectionality helps to deconstruct these layers, demonstrating how gender-based harassment and connectivity hurdles interact to perpetuate existing inequalities, eventually limiting who can fully participate in digital feminist movements.

Discussion

The interviews provide a comprehensive exploration of digital feminism and activism in contemporary contexts, revealing several key insights. They highlight the transformative role of the digital sphere in amplifying women's voices and addressing gender inequalities in Mozambique. Participants emphasized that digital platforms offer a vital space for raising awareness about women's rights and mobilizing support, particularly in contexts where traditional media may be inaccessible or ineffective. This underscores the significance of digital activism in democratizing access to advocacy and enabling marginalized voices to be heard, mainly for women (O'Byrne, 2019).

A significant finding from the interviews is the multifaceted challenges faced by women in the digital realm. Digital violence emerged as a pervasive issue, with participants noting instances of harassment, exposure and intimidation aimed at silencing feminist voices. These attacks not only undermine women's efforts but also perpetuate existing power dynamics and societal prejudices. Moreover, the high cost of internet access poses a barrier, particularly in rural areas, limiting women's participation in online activism and advocacy efforts. This is not only the case in Mozambique, as it was demonstrated that in countries like Nigeria and Ghana, such challenges prevent women from doing their activism online (Chiluwa, 2024).

Another critical insight from the interviews is the intersectional nature of digital feminism. Participants highlighted that while digital platforms facilitate solidarity among women, they also reflect and sometimes perpetuate broader societal inequalities. Issues such as cultural criticisms, traditional beliefs and patriarchal norms were identified as barriers to the widespread acceptance and effectiveness of digital feminist campaigns. This complexity necessitates nuanced approaches that acknowledge and address diverse experiences and identities within feminist movements. It echoes the analysis made by McArthur (2016), who demonstrated that despite the largely degrading media representations of Blackness, historically, Black girls and women have been strong activists, disrupting narratives the media conveys about Black girlhood and womanhood.

Additionally, it was noted that the interviews reveal tensions within digital feminism regarding inclusivity and representation. Participants discussed the challenges of ensuring that digital feminist spaces are inclusive of all women, particularly those from marginalized communities who may face intersecting forms of discrimination. They emphasized the importance of humility in digital feminist advocacy, advocating for approaches that prioritize listening to and amplifying diverse voices rather than imposing centralized perspectives.

This situation underscored the evolving nature of digital feminism and the ongoing need for strategic adaptation. Participants noted that while digital activism has expanded opportunities for women's empowerment and advocacy, it remains a contested space where gains are continually challenged – this refers to the digital divide as a main challenge (Vachhani, 2023). Interviewees highlighted the importance of resilience and solidarity among women activists in navigating these challenges and advancing collective goals of gender equality and social justice in both digital and physical spheres.

Conclusion

This chapter represents a first scientific attempt to approach feminism from a virtual point of view, so as authors we assume some limitations in its conclusions. Having said that, it is important to emphasize that this chapter discussed the transformative role of digital platforms in amplifying women's voices and advancing gender equality, with Mozambique as a case study. It revealed how digital spaces provide vital avenues for raising awareness about women's rights and mobilizing support, particularly in contexts where traditional media may fall short. However, it also highlighted persistent challenges, such as digital violence and unequal access to the internet, which hinder the full realization of digital feminist goals in Mozambique and beyond.

Furthermore, using two case studies, this chapter revealed the intersectional complexities within digital feminism, noting how cultural norms and societal inequalities shape the effectiveness of feminist advocacy online. It emphasized the importance of inclusive digital spaces that accommodate diverse perspectives and experiences, especially for marginalized women facing multiple forms of discrimination. It also shows how the emerging digital collectives fuse themselves between the online – as it is the case of the Ntavase campaign – and the offline Observatório das Mulheres arenas. The chapter concludes by advocating for continued dialogue, strategic adaptation and collective resilience among women activists to confront these challenges and foster lasting social change.

Bibliography

Agarwal, B., (1988), *Structure of Patriarchy*, New Delhi: Kali for Women.

Aguiar, N. (2000), Patriarcado, Sociedade e Patrimonialismo, *Sociedade & Estado*, 15 (2): 303-30.

Arnfred, S. (2011), *Sexuality and Gender Politics in Mozambique: Rethinking Gender in Africa*, Rochester, NY: Boydell & Brewer.

Bonate, L. and Katto, J. (2021), Women in Mozambique, in *The Oxford Research Encyclopedia of African History*, New York: Oxford.

Casimiro, I. (2001) 'Repensando as Relações Entre Mulher e Homem no Tempo de Samora', in *Samora Homem do Povo*, Maguezo Editores, Maputo, 99-107.

Casimiro, I. (2004), 'Paz Na Terra, Guerra em Casa'. Feminismo e Organizações de Mulheres em Moçambique, Maputo, Edição PROMÉDIA, Colecção Identidades.

Casimiro, I. (2014), 'Paz Na Terra, Guerra em Casa'. Feminismo e Organizações de Mulheres em Moçambique, Série Brasil & África, Coleção Pesquisas 1, Editora UFPE, Recife.

Chiluwa, I. (2024), Discourse, Digitisation and Women's Rights Groups in Nigeria and Ghana: Online Campaigns for Political Inclusion and Against Violence on Women and Girls, *New Media & Society*, January 2024.

Choi, M. and Cristol, D. (2021), Digital Citizenship with Intersectionality Lens: Towards Participatory Democracy Driven Digital Citizenship Education, *Theory into Practice*, 60 (1).

Cortesão, I. and Casimiro, C. (2021), Twelve Mozambican Feminisms: Between the Local and the Global, in A. Tambe and M. Thayer, (eds) *Transnational Feminist Itineraries: Situating Theory and Activist Practice*, New York: Duke University Press, pp. 207-21.

Coward, R. (1983), *Patriarchal Precedents: Sexuality and Social Relations*, London: Routledge.

Crenshaw, K. (2004), A Interseccionalidade na Discriminação de Raça e Género, in VV. AA., ed. *Cruzamento: Raça e Género*, Brasília: Unifem.

DataReportal (2025), Digital 2025: Mozambique, 3 March, Available at https://datareportal.com/reports/digital-2025-mozambique.

Disney, J. L. (2008), *Women's Activism and Feminist Agency in Mozambique and Nicaragua*, Philadelphia, PA: Temple University Press.

Eisenstein, Z. R. (1979), *Capitalist Patriarchy and the Case for Socialist Feminism*, New York: Monthly Review Press.

Foucault, M. (1991), *Discipline and Punish: The Birth of a Prison*, London: Penguin.

Freire, P. (1983), *Education for Critical Consciousness*, New York: Seabury Press.

Gaventa, J. (2003), Power After Lukes: A Review of the Literature, Brighton: Institute of Development Studies.

Haraway, D. (1988), Situated Knowledges: The Science Question in Feminism and the Privilege of Partial Perspective, *Feminist Studies*, 14 (3): 575-99.

Haraway, D. (2004), 'Género' Para um Dicionário Marxista: A Política Sexual de Uma Palavra, *Cadernos Pagu*, June 2004.

hooks, b. (1981), *Ain't I a Woman: Black Women and Feminism*, Cambridge: South End Press.

Joanguete, C. and Tsandzana, D. (2023), *Cidadania Digital*, Maputo: Ethale.

Karberg, S. (2015), Female Political Participation and their Influence Towards Greater Empowerment of Women in Mozambique, Friedrich Ebert Stiftung.

Koutsogiannis, D. and Adampa, V. (2012), Girls, Identities and Agency in Adolescents' Digital Literacy Practices, *Journal of Writing Research*, 3 (3): 217-47.

Kozinets, R. V. (2019), *Netnography: The Essential Guide to Qualitative Social Media Research*, 3rd ed, London: Sage Publications.

Lilja, M. (2022), The Definition of Resistance, *Journal of Political Power*, 15 (2): 202-20. Available at: https://doi.org/10.1080/2158379X.2022.2061127 [Accessed 3 September 2024].

McArthur, S. (2016), Black Girls and Critical Media Literacy for Social Activism, *English Education*, 48 (4): 362-79.

Mpofu, S. (2016), 'Digital Activism in Social Media Era: Critical Reflections on Emerging Trends in Sub-Saharan Africa', in *Blogging, Feminism and the Politics of Participation: The Case of Her Zimbabwe*, London: Palgrave Macmillan.

O'Byrne, W. I. (2019), Educate, Empower, Advocate: Amplifying Marginalized Voices in a Digital Society, *Contemporary Issues in Technology and Teacher Education*, 19 (4): 640-69.

Osório, C. (2010), *Género e Democracia; As eleições de 2009 em Moçambique*, WLSA, Maputo.

Oyedeme, T. (2020), 'Handbook of Communication for Development and Social Change', The Theory of Digital Citizenship, Singapore: Springer.

Paasonen, S. (2011), Revisiting Cyberfeminism, *Communications*, 36 (3): 335-52.

Pateman, C. (1988), *The Sexual Contract*. Stanford, California: Stanford University Press, (Cited in: Aguiar, N., 2022).

Roberts, T. and Bosch, T. eds (2023), *Digital Citizenship in Africa: Technologies of Agency and Repression*, London: Zed Books.

Rubin, H. J. and Rubin, I. S. (2012), *Qualitative Interviewing: The Art of Hearing Data*, 3rd ed., London: Sage Publications.

Tripp, A. (2001), Women's Movements and Challenges to Neopatrimonial Rule: Preliminary Observations from Africa, *Development and Change*, 32 (1): 33-54.

Tsandzana, D. (2018), Urban Youth and Social Networks in Mozambique: The Political Participation of the Helpless Connected, *Sociedade e Comunicação*, 34 (2): 251-65.

Tsandzana, D. (2020), Redes Sociais da Internet Como 'Tubo de Escape' Juvenil no Espaço Político-Urbano em Moçambique, *Cadernos de Estudos Africanos*, 40 (2): 167-89.

Vachhani, S. J. (2023), Networked Feminism in a Digital Age – Mobilising Vulnerability and Reconfiguring Feminist Politics in Digital Activism, *Gender, Work and Organization*, 31 (3): 1031-48.

Chapter 9

DIGITAL FEMINIST CITIZENSHIP IN MALAWI: MWIZA CHAVURA'S RAPE SONG

Jones Maweranga and Godwins Lwinga

This chapter presents a discussion on the phenomenon of digital feminist citizenship in Malawi: a case study of Mwiza Chavura's rape song. The study aims to discuss the dynamics of digital feminist citizenship as exemplified in the case study of Mwiza Chavura's rape song.

The chapter pursues the following main research question about digital feminist citizenship in Malawi: How has the use of digital platforms in Malawi enhanced digital feminist citizenship and enabled collective agency in resistance to rape culture? In line with the main research question, the following two research sub-questions are also addressed: (1) What is the significance of Mwiza Chavura's rape song about digital feminist citizenship in Malawi? (2) What forms of gender injustice do Malawian feminists raise on various digital platforms?

Mendes, Ringrose and Keller (2019) define feminist digital citizenship as women's employment of digital technologies to claim their rights as citizens, challenging sexism, rape culture and structures that hamper their active participation in society. Digital feminist citizenship in Malawi is significant because it enables women to mobilize themselves against GBV. For instance, Malawian women digitally mobilized themselves to protest against Mwiza Chavura's song entitled '*Ndidzakupanga Rape*', which can be translated as 'I will rape you'.

The chapter is guided by the conceptual framework of cyberfeminism, which denotes the feminist appropriation of digital citizenship. Kennedy (2000) traces the development of cyberfeminism in relation to its origin in feminist theory and praxis during the late 1980s and early 1990s. In its initial evolution, cyberfeminism engaged with the emergent ICT revolution, subsequently becoming one of the catalysts in the digital feminist citizenship discourses (Baer, 2018). Kember (2002) gives a fivefold motivation for the rise of cyberfeminism, which also serves as a justification for its adoption by this study as a conceptual framework. First, it was a reaction to the radical politics of cyberpunk. Second, it was a rebellion against the cyber-dominance of middle-class adolescent males by destabilizing patriarchal culture and strategically using essentialist images of women. Thus, cyberfeminism is essentially concerned with neutralizing men's dominance in the digital citizenship sphere. Third, cyberfeminism allows women users to engage in digital citizenship by sharing information, participating in recreation and having discussions in various digital platforms. Consequently, Kuah-Pearce (2008) explains that 'the information galaxy, the cyberspace and the Internet ... are no longer viewed as a masculine space and tool as women have not only embraced but also used the cyberspace to negotiate and reframe themselves within existing social structure' (276). Fourth, cyberfeminism offers feminist agency by empowering women and girls to engage in purposeful action and pursue their own goals and interests, free from the threat of violence and retribution of any kind. Fifth, cyberfeminism provides a platform for feminist resistance by allowing women to appropriate their rights and engage in digital feminist activism against gender inequalities and inequities, GBV, etc.

Therefore, in pursuit of the main research question, the chapter is divided into the following eight sections: (1) introduction, (2) background, (3) digital feminist citizenship in Malawi, (4) literature review, (5) analysis, (6) methodology, (7) discussion of research findings and (8) conclusion and recommendations.

Background

In this section of the chapter, we provide a background of Malawi in three different historical periods: (1) 1860–1964, (2) 1964–1994 and (3) 1994–2024.

The first period, from 1860 to 1964, is characterized by women who played a significant role in liberating Malawi from colonial rule. During this time, their rights claims were primarily about national liberation, and women were involved in political mobilization and agency through political leadership, street demonstrations and military struggle. Banda and Kayira (2012) explain that women actively fought alongside men in the nationalist liberation wars, and some of them died as martyrs. Also, women were detained alongside men – for example, Rose Lomathinda Chibambo, Vera Chirwa and Gertrude Rubadiri. The predominant forms of media used by women at this time were radio (Federal Broadcasting Corporation and Radio Malawi) and newspapers (*Nyasaland Times, Odini*, etc.).

The second period, from 1964 to 1994, is characterized by Malawi's independence from British colonial rule and the era of Dr Hastings Kamuzu Banda, the first president of Malawi. Semu (2002) argues that Dr Banda's dictatorship manipulated the status of women (*Mbumba za Kamuzu*) to sing political songs and dance for him while praising his heroism, leadership style and victory over colonialism and indigenous rebels (*Zigawenga*). Chirwa (2001) argues that women had to negotiate daily with the regime's oppressive machinery, which abused them through the songs and dances, subsequently exposing them to sexual exploitation by politicians and the youth militia. The predominant forms of media employed by women at this time were radio (Malawi Broadcasting Corporation), newspapers (*Malawi News, Daily Times, Boma Lathu*, etc.) and magazines (*Moni, Police Magazine*, etc.).

Third, the period from 1994 to 2024 is characterized by the democratic dispensation and the implementation of the digitalization project, which has enabled the praxis of feminist digital citizenship in Malawi. Kemp (2024), in *Digital 2024: Malawi*, gives a threefold state of digital use in Malawi. First, Malawi had 5.86 million internet users at the beginning of 2024. Meanwhile, internet penetration stood at 27.7 per cent. Second, 1.3 million Malawians actively used social media in January 2024, equating to 6.1 per cent of the total population. Third, a total of 11.77 million cellular mobile connections were active in Malawi in early 2024, which is equivalent to 55.5 per cent of the total population. Nevertheless, despite Malawi registering low levels of digitalization, women have employed various digital media, such as WhatsApp, Facebook, Twitter, Instagram, LinkedIn and YouTube, and online tabloids, such as *Nyasa Times, Maravi Post* and *Malawi24* to demonstrate their engagement with digital citizenry. Consequently, this has enabled women to engage in online organization and raise issues ignored by mainstream media, political parties and society in general.

Digital feminist citizenship in Malawi: A case study of Mwiza Chavura's 'Ndidzakupanga Rape' song

Pasula (2018) reports that when Mwiza Chavura released his song entitled 'Ndidzakupanga Rape' (I will rape you) in 2018, there was a backlash because it was deemed as inciting sexual violence against women in Malawi. In the song, Chavura is announcing a plot that involves getting a girl drunk on his bills to rape her simply because she had turned down his sexual advances. The song's background vocals depict a woman crying for help in the act of rape.

Mwiza Chavura's *'Ndidzakupanga Rape'* song is significant because it precipitated an unprecedented wave of digital feminist citizenship in Malawi whereby women self-mobilized to condemn the song and resist a deep-rooted rape culture. Consequently, most women employed social media, especially Facebook, and online tabloids, such as *Nyasa Times, Maravi Post,* and *Malawi24*, to challenge and condemn both the artist and the song. Pasula (2018) narrates that several women denounced the song. Emma Kaliya described it as demeaning to womenfolk and decried the glorification of rape. Esmie Tembenu strongly condemned the song because it represented a form of sexual violence against girls and women in Malawi by depicting sexual harassment through rude jokes and suggestive remarks. She further rebuked some Malawians who celebrated and defended the rape song in the name of freedom of expression by pointing out that although the Malawi Constitution guarantees freedom of expression, it has to be exercised responsibly to avoid infringing on women's rights. Dumase Mapemba described the rape song as being retrogressive for advocating rape and insulting the modesty of a woman, which is contrary to Section 137(3) of the Penal Code. Martha Mituka described Chavura as a disgrace and called for Mwiza Chavura's apology and subsequent banning of the song (Pasula, 2018).

A network of Christian artists in Malawi raised a threefold objection to the *'Ndidzakupanga Rape'* song. First, it expressed sadness over the song's level of vulgarity and condemned the trivializing of rape and other forms of sexual violence, which violates women's rights and well-being. Second, it condemned some online music distribution sites such as www.malawi-music.com and @entertainmentmalawi7687 for making the song available. Third, it called for the song's banning and the booking of its creators, including the producers, the label Nyambaro Music, under which the song was produced, and all websites distributing it (Pasula, 2018).

The digital feminist activism against Mwiza Chavura's *'Ndidzakupanga Rape'* song yielded two positive results. It eventually led to Mwiza Chavura's arrest and subsequent charge for producing obscene material contrary to Section 179(1)(a) of the Penal Code. It also led to the banning of the song. Miss Anganile Nthakomwa, an officer of the Malawi Censorship Board, announced that the board had prohibited the song after thoroughly examining its content using the provisions of Section 23 of the Censorship and Control of Entertainment Act. She stated that 'In our considered opinion, the song mocks the modesty of women, seems to be celebrating sexual violence. The song has the potential of influencing the perception of the general public on sexual violence against women, reduces women to sex objects, and offending those who have been victims of rape and other forms of sexual violence' (Masina, 2018: 1).

Literature review

In this section of the chapter, we present a literature review pertinent to the topic 'Digital feminist citizenship in Malawi: A case study of Mwiza Chavura's rape song'. The literature review is divided into two parts: (a) digital citizenship and digital feminist citizenship and (b) rape culture.

Digital citizenship

In this article, 'digital citizenship' is defined according to the insights offered by Bosch and Roberts (2023), who identify six different aspects of digital citizenship. First, digital citizenship constitutes participating in civic life by employing digital tools or navigating online spaces. Second, various digital tools and online platforms afford citizens limitless opportunities for civic engagement. Third, digital citizenship is not given by the state but is actualized when individuals employ mobile and internet technologies in their spiritual, socio-economic, political and cultural relations in partaking in digital citizenship. Fourth, for an individual to engage in digital citizenship, two prerequisites are necessary: access to digital tools and digital literacy, which refers to the capacity to use the various digital technologies. Fifth, the state provides a conducive legal environment for enacting laws that regulate the digitalization industry. Sixth, private corporations have a role in ensuring connectivity, exercising the power to increase or limit digital citizenship.

The article further defines 'feminist digital citizenship' in line with Lister (2017), who asserts that 'the feminist digital citizenship should be marked by theoretical and practical considerations that give due credit to women's self-mobilisation and agency rather than simply seeing women as victims of discrimination and oppressive male-dominated political, economic, and social institutions' (6). Although the body of knowledge concerning digital citizenship in Africa has advanced (Bosch and Roberts, 2023; Ogbonna, 2018), little research exists that streamlines digital feminist citizenship in Africa. Therefore, this significant knowledge gap necessitates the generation of literature that engages with the following question: How has the use of digital platforms in Malawi enhanced digital feminist citizenship and enabled collective agency in resistance to rape culture? The expectation is that the extensive use of digitalization mechanisms, particularly in Malawi, is supposed to reshape and influence the nature of digital feminist citizenship. Consequently, this article contributes to the existing body of knowledge on digital feminist citizenship in Malawi by interrogating the intersectionalities of Mwiza Chavura's rape song and the praxis of digital feminist citizenship in the country.

Yun (2020) observes that digital citizenship was once hegemonic, characterized by macho network patterns, cybersexism, online misogyny and gendered cyberhate, but now feminist praxis, such as the #MeToo movement and protests against femicide and other forms of GBV, have transformed the digital space into feminist resistance. In particular, the #MeToo movement highlights women's self-mobilization and agency in resisting a widespread and established rape culture that women have to overcome on a daily basis in both physical and cyber spaces. Moreover, the inadequacy of the legal framework to resolve the various ways that unconscious sexist bias continues to proliferate the public sphere and influence GBV, gender inequities and inequalities in the twenty-first century were underscored by the #MeToo movement (Cavalieri, 2019).

Clark-Parsons (2021) argues that hashtag feminism, which deploys Twitter's (X) metadata tags in organizing posts aimed at drawing visibility to a feminist cause, demonstrates the praxis of digital feminist citizenship. Tarana Burke is credited for launching the Me Too movement in 2006 as a means of generating feminist empowerment through empathy for young women of colour who had survived sexual violence and were suffering in silence (Rodino-Colocino, 2018). A decade later, the hashtag #MeToo was popularized by Alyssa Milano, a Hollywood actor, in reaction to sexual abuse allegations against renowned producer Harvey Weinstein. Alyssa Milano's invitation to her Twitter followers to write 'me too' if they had ever experienced any form of sexual violence precipitated an unprecedented wave of digital feminist citizenship by creating an outlet for survivors of various forms of sexual violence to narrate their stories, thereby sparking a global conversation as far as sexual harassment and assault are concerned (Clark-Parsons, 2021). Consequently, hashtag feminist praxis heightened women's engagement with digital feminist citizenship by amplifying feminist voices, self-mobilization, agency, empowerment and solidarity (Ryland, 2018).

Yun (2020) observes the twofold stages of digital feminism, which are delineated by the Web 1.0 era and the Web 2.0 era. On one hand, the Web 1.0 era was characterized by a distinction between online and offline spaces, and internet users were considered to be content consumers generated by the servers. On the other hand, the Web 2.0 era is characterized by hyper-connectivity, which enables an intersection of online and offline spaces and allows internet users to be both content creators and consumers, thus creating an ambience for a participative social Web and reshaping the contemporary ontological digital condition with the following axiom: 'I am connected to the social media; hence, I exist' (48). In this respect, women have leveraged their digital connectivity to denounce sexism daily, expose the foolishness of misogynistic culture and disrupt rape cultural thinking and practices. Moreover, digital feminism has bridged the gap that existed between offline and online feminist struggle and resistance (Fotopoulou, 2016).

Ndasauka (2021) suggests four reasons that motivate women to engage in digital feminist citizenship in Malawi. First, the human need for community and communion based on the African communitarianism enables women to engage in digital feminist citizenship as a socializing agent with the potential for facilitating both offline and online social engagement. Second, digital feminist citizenship has enabled women in Malawi to challenge deeply rooted and prevailing sociocultural and religious gender stereotypes, gender inequalities and inequities (Tembo, 2024). Third, digital feminist citizenship has enabled women in Malawi to be connected to sexual and reproductive health rights (Klason, 2016). Fourth, women in Malawi are engaging in digital feminist citizenship either positively to build each other up or negatively to tear each other down. This was heightened after the botched 2019 presidential election, which was chaired by Jane Ansah, the former chairperson of the Malawi Electoral Commission. Two significant issues transpired. First, Malawians associated with the opposition parties participated in violent demonstrations calling for Jane Ansah's resignation. Second, women in Malawi were divided on whether to support or denounce Jane Ansah, and many used various digital platforms to express their views (Chunga, 2020).

Pasula (2018) argues that women's engagement in digital feminist citizenship in Malawi was heightened through their self-mobilization and agency in resisting a rape culture that was deemed to be amplified by Mwiza Chavura's *'Ndidzakupanga Rape'* song. Most women employed various digital platforms, such as Facebook and online tabloids, to condemn Mwiza Chavura's song. Some directly engaged with him on his Facebook page, demanding an apology, to which he complied. In this respect, the praxis of digital feminist citizenship eventually led to the banning of the song and the raising of digital consciousness in fighting rape and other forms of GBV in Malawi (Masina, 2018).

Rape culture

Powell (2015) defines rape culture as a social context in which 'sexual violence against women is implicitly and explicitly condoned, excused, tolerated and normalised' (575). Buchwald, Fletcher and Roth (2005) contend that a rape culture refers to 'a complex set of beliefs that encourage male sexual aggression and support violence against women' (11).

Garvey (2005) traces the historical development of the concept of rape culture to the 1970s, when feminists delivered a radical critique of the traditional and conventional assumptions of rape as an abnormal behaviour of a deviant person. Rather, the feminists argued that rape was associated with and facilitated by various psycho-social and cultural factors. Edwards (1987) proposes that the pervasiveness of rape culture is partly due to the celebration of male dominance in society. Ringrose et al. (2021) argue that rape as male sexual aggression against women is precipitated by two conditions that abdicate responsibility on the part of the man and vilify womanhood. The first is victim-blaming, a scenario in which the woman is blamed for causing the rape incident, and this is further reflected in conversations that minimize the severity of rape and excuse the man who perpetrated such a horrendous crime. Second, rape culture trivializes certain aspects of heterosexuality to the extent of normalizing and naturalizing male sexual aggression and female impassiveness. Surprisingly, the culture of rape is widely tolerated and accepted in many societies, while rape itself is condemned.

According to Mkhize et al. (2022), rape culture is intensified by the use of misogynistic language, the sexual objectification of women's bodies and the fascination with sexual violence, thereby creating a society that holds women in disdain, demeans their ontological constitution, disrespects their rights and disregards their safety and well-being. Unfortunately, the rape culture exacerbates women's vulnerability due to the prevalence of sexual violence (Dlakavu, 2016).

Mkhize, Khanyisile and Olofinbiyi (2022) argue that rape culture is associated with beliefs and practices that normalize sexual violence by exonerating the perpetrator and blaming the victim. These insinuations are prevalent in both social media and mainstream media and constitute discussions that in certain ways justify rape by blaming women for being the causative factor.

Gaskin (2019) asserts that the concept of rape culture should not be limited to incidents where women are raped but should extend to men being raped, as well as members of the lesbian, gay, bisexual, transgender, queer and intersex (LGBTQI) communities becoming victims of sexual violence in what is commonly called 'corrective rape', a phenomenon prevalent in South Africa (Mawerenga, 2018). Thus, rape culture constitutes an intersectionality of various factors such as gender, race, ethnicity, sexual orientation, disability and power relations (Sande and Chirongoma, 2021). Brown (2018) gives a fourfold description of the characteristics of male sexual aggression in the form of rape perpetrated against women. First, there is the demonization of victims. Second, downplaying the seriousness of the crime. Third and fourth, sexualizing and normalizing sex and rape in the media.

Sande and Chirongoma (2021) identify three factors that are responsible for fuelling rape culture across the world: 'hegemonic masculinity, rape myths, and language' (3). Rape culture is further engrained by the conjecture that male sexual aggression is the manifestation of masculine characteristics that enable men to take a dominating and leading role in exercising their position of power and authority over women.

Najumi (2013) states that rape culture is bolstered by misogynistic language, which is employed in a teasing way or through jokes that trivialize and justify rape. For instance, Mwiza Chavura's *'Ndidzakupanga Rape'* song downplays the gravity of rape and celebrates the victim's physical and psycho-social pain as a result of rape. Moreover, rape is considered as a justified form of revenge to a girl who does not consent to a man's sexual advances (Pasula, 2018).

According to Fraser (2015), rape myths play a crucial role in protecting rape culture because they shape attitudes concerning rape and offer explanations of people's behaviour towards rape victims and perpetrators. Burnett et al. (2009) list four statements that propagate rape myths. First, 'women say "NO" to sex when they mean "Yes."' Second, 'any woman who gets raped is promiscuous'. Third, 'women falsify their rape when they had consented to having sex.' Fourth, 'it's the women's seductive attire that aroused the man to rape her'. Mkhize, Khanyisile and Olofinbiyi (2022) mention the following mythical statements concerning rape. First, 'husbands cannot rape their wives.' Second, 'women enjoy rape'. Third, 'women ask to be raped' (8–13).

Sande and Chirongoma (2021) contend that the deconstruction of a rape culture entails an understanding of the role played by the male sex organ, the penis, as a tool of rape in a heterosexual sex activity. Thus, rape culture goes beyond just sexual coercion and concerns penetration of the penis into the vagina. Ngubane-Mokiwa and Chisale (2019) concur that 'the most common sexual penetration among the heterosexual is culturally and religiously constructed as the penetrator who is regarded as the powerful on top and the penetrated and powerless under' (109).

Analysis

The chapter analyses the following digital platforms: Facebook and online tabloids, such as *Nyasa Times*, *Maravi Post*, *Malawi24*, and *Pan African Visions*. These platforms were chosen because they facilitated digital feminist citizenship in Malawi by enabling women to challenge a rape culture, that was heightened by Mwiza Chavura's *'Ndidzakupanga Rape'* song.

The *'Ndidzakupanga Rape'* song enhanced digital feminist citizenship in Malawi by triggering numerous feminist reactions, particularly from women who challenged the promulgation of rape culture. For instance, Anganile Nthakomwa, Martha Mituka, Emma Kaliya, Dumase Mapemba and Esmie Tembenu denounced the song. These are prominent Malawian women known for their respective contributions to Malawian society. Apart from these women, numerous women engaged in digital feminist citizenship by condemning Mwiza Chavura's song on his Facebook page and other Facebook pages.

The data for this study was qualitative in nature and constituted specific responses given by various women that demonstrated their digital feminist citizenship through their engagement with the *'Ndidzakupanga Rape'* song.

Data selection for this chapter was guided by the feminist thematic data analysis approach. Thus, key factors of feminist data analysis were employed. In this respect, the crucial question was 'How did the *"Ndidzakupanga Rape"* song entrench patriarchy, discrimination, sexual objectification, oppression and stereotyping against women?' (Carrino et al., 2022). Thus, feminist thematic data analysis was employed to draw concepts emerging from the data, and linkages were made between themes and patterns, which were finally interpreted accordingly.

The nature of women's responses to the *'Ndidzakupanga Rape'* song is feminist because they captured women's experiences, perspectives, voices, narratives and actions in challenging rape culture. Moreover, the women's responses were aimed at uncovering and challenging gender biases, inequalities and power imbalances within the context of the *'Ndidzakupanga Rape'* song (Whaley, 2001).

Methodology

The chapter employed qualitative research methods, such as a literature review and feminist narratology. First, the authors conducted a comprehensive literature review on the topic: 'Digital feminist citizenship in Malawi: A case study of Mwiza Chavura's rape song'. The literature review was delimited by the chapter's aim, namely, to discuss the dynamics of digital feminist citizenship as exemplified in the case study of Mwiza Chavura's rape song.

The literature review was useful for two reasons. It enabled the authors to underpin the discussion with relevant feminist theoretical and conceptual frameworks. It was also useful in identifying knowledge gaps that could be filled by the present study.

The chapter employed African feminist narratology because it allows women to tell their woman-centric stories in seven ways. First, the stories become a source of women's knowledge production and distribution. Second, stories are an alternative method of capturing women's knowledge and perspectives. Third, stories describe how women make sense of their world. Fourth, stories are a vehicle for exploring women's lived experiences and realities. Fifth, stories allow for the contextualization of women's lived experiences and realities away from dominant and traditional methods. Sixth, stories are employed as tools for feminist resistance against patriarchal control, manipulation, exclusion and oppression of women in African society. Seventh, stories are a resource for framing women's self-mobilization and agency in a liberative, restorative and empowering manner to African women's psycho-social well-being (Sheik, 2018).

The theory and praxis of African feminist narratology are both significant and relevant for this chapter because they address research questions in the following four ways. First, African feminist narratology enables women who are traditionally marginalized and disempowered to reclaim their voices. Second, through the appropriation of personal narratives, oppressed women are empowered to create their own sphere of existence, thereby attaining self-emancipation from their previous marginalized positions in society. Third, it positions women's lived experiences and realities at the centre and facilitates feminist resistance to rape culture. Fourth, it provides a platform for sharing woman-centric stories, thereby precipitating women's self-mobilization, creation of networks and agency (Amoah, 1997).

Discussion of findings

The chapter interrogated the phenomenon of digital feminist citizenship in Malawi: a case study of Mwiza Chavura's rape song. The aim of the study was achieved by analysing three thematic areas: (1) the praxis of digital feminist citizenship in Malawi, (2) the significance of Mwiza Chavura's song in relation to digital feminist citizenship and (3) forms of gender injustice raised by Malawian feminists on various digital platforms.

The praxis of digital feminist citizenship in Malawi

Study findings revealed that the praxis of digital feminist citizenship in Malawi has been appropriated by women in the country, contrary to the findings of Kambandu and Manduwi (2022), who argued that digital citizenship is perceived to better suit men than women. Moreover, Malawian women have navigated various obstacles that prevent them from employing digital feminist citizenship. These include the socio-cultural, economic and political factors that compound their full digital citizenship participation (Malera and Chisala-Tempelhof, 2018).

Study findings established that the praxis of digital feminist citizenship in Malawi in connection with Mwiza Chavura's rape song was manifested through social media, especially Facebook, and online tabloids, such as *Nyasa Times*, *Maravi Post* and *Malawi24*, to challenge and condemn both the artist and the song. Through an active engagement with digital feminist citizenship, Malawian women heightened the feminist consciousness in the country concerning women's safety, well-being, rights and resistance to rape culture (Pasula, 2018).

The praxis of digital feminist citizenship in Malawi highlighted the need for ensuring that the creative industry does not overstep its freedom of expression to the extent of violating the freedom and rights of women (Pasula, 2018). Malawian women were in the forefront of confronting, challenging and condemning both Mwiza Chavura and his song; hence, they exercised their digital feminist citizenship in resisting rape culture, affirming their dignity and reclaiming their rights (Masina, 2018). Thus, the use of digital platforms in Malawi enhanced digital feminist citizenship and enabled collective agency in resistance to rape culture.

The significance of Mwiza Chavura's song to digital feminist citizenship

Study findings revealed that Mwiza Chavura's *'Ndidzakupanga Rape'* song fostered a rape culture in Malawi by propagating misogyny, sexual violence and rape myths. The *'Ndidzakupanga Rape'* song can properly be described as being misogynistic. Adams and Fuller (2006) define 'misogyny as the hatred or disdain of women. It is an ideology that reduces women to objects for men's ownership, use, or abuse. This ideology is widespread and common throughout society' (939).

Mwiza Chavura's song was a demonstration of the use of demeaning and hateful language against women (Pasula, 2018). The lyrics projected a notion of sexual objectification, in which a woman's body or body parts are separated from her and reduced to mere instruments or regarded as if they represent the entire person. The lyrics further demonstrated how language is used to construct womanhood and femininity to be in servitude to patriarchy. For instance, the composition of '*Ndidzakupanga Rape*' constitutes both an expression of ideas and a way of practically living them. In other words, a song's lyrics depict hegemonic masculinity, misogyny and dominant ideologies concerning gender and sexuality. In this case, the '*Ndidzakupanga Rape*' song becomes a socializing agent of society (Chiweshe and Bhatasara, 2013).

Study findings indicate that Mwiza Chavura's misogynistic and sexist song met unprecedented resistance, especially through the praxis of digital feminist citizenship by Malawian women. Women employed Facebook and online tabloids, such as *Nyasa Times*, *Maravi Post* and *Malawi24*, to challenge and condemn both the artist and his '*Ndidzakupanga Rape*' song.

Forms of gender injustice raised by Malawian feminists on various digital platforms

Study findings revealed various forms of gender injustice raised by Malawian women on different digital platforms. Malawian feminists have engaged with cyberviolence against women on various digital platforms. Women experience traumatic forms of cyberviolence, such as cyber-stalking, cyber-harassment, cyber-bullying, online sexual harassment, nonconsensual pornography and body-shaming (Malanga, 2021).

Malawian feminists have contended with political violence directed towards women to instil fear and prohibit their political participation. For instance, a WhatsApp video that went viral on 20 January 2019 depicts Veronica Katanga, a United Transformation Movement supporter, being stripped of her party regalia, beaten and verbally harassed. 'Her bra was torn, and she was left standing by the roadside with one breast exposed. The attackers took her phone and filmed the entire event. They also stole her money' (Chisala, 2019). Thus, Malawian feminists employed the various digital media platforms that circulated the WhatsApp video to engage with politically motivated GBV.

Malawian feminists have engaged with the issue of intimate partner violence on various digital platforms. For instance, a husband chopped off his wife's hands and a leg after a family dispute in Dowa district. Unfortunately, the woman was pronounced dead upon arrival at the hospital due to loss of blood. This incident elicited unprecedented responses from Malawian feminists employing various digital platforms to curb the rising cases of intimate partner violence in the country.

Conclusion and recommendations

The chapter has presented a discussion on the phenomenon of digital feminist citizenship in Malawi: a case study of Mwiza Chavura's rape song. It has highlighted the multifaceted dynamics of digital feminist citizenship as exemplified by Malawian women in challenging and condemning the propagation of rape culture, which was heightened by the '*Ndidzakupanga Rape*' song. Moreover, it has demonstrated Malawian women's self-mobilization and agency in employing various digital platforms to engage with different forms of gender injustice.

The chapter makes the following three recommendations. The first is the need to empower Malawian women with the knowledge, use and ownership of digital media platforms so that they can effectively exercise their digital feminist citizenship to advance the feminist cause. Second, the praxis of digital feminist citizenship by Malawian feminists has the potential to amplify the voices of the voiceless and establish connections between the voices as a means of fighting rape culture. Third, Malawian women should go beyond the employment of online tabloids and Facebook to utilize other digital platforms, such as X. In this case, they can learn from the successes of hashtag feminist activism, such as #MeToo.

Bibliography

Adams, T. M. and Fuller, D. B. (2006), The Words Have Changed but the Ideology Remains the Same: Misogynistic Lyrics in Rap Music, *Journal of Black Studies*, 36 (6): 938–57.

Adeniyi, E. (2022), Penis Primacy, Female Marginality, and Masculine Subjectivity in 'Nigeria's Rape Culture', *Quarterly Review of Film and Video*, 39 (6): 1308–36.

Amoah, J. (1997), Narrative: The Road to Black Feminist Theory, *Berkeley Women's LJ*, 12, 84.

Baer, H. (2018), Redoing Feminism: Digital Activism, Body Politics, and Neoliberalism, in *Digital Feminisms* (pp. 25–42), London and New York: Routledge.

Banda, P. C. and Kayira, G. W. (2012), The 1959 State of Emergency in Nyasaland: Process and Political Implications, *The Society of Malawi Journal*, 65 (2): 1–19.

Brown, S. (2018), *Rape Culture or a Culture of Rape? American Rape Culture Compared to South African Rape Accommodating Culture*, Doctoral dissertation.

Buchwald, E., Fletcher, P. R. and Roth, M. (2005), *Transforming a Rape Culture* (2nd ed.), Minneapolis, MN: Milkweed Editions.

Burnett, A., Mattern, J. L., Herakova, L. L., Kahl Jr D. H., Tobola, C. and Bornsen, S. E. (2009), Communicating/Muting Date Rape: A Co-Cultural Theoretical Analysis of Communication Factors Related to Rape Culture on a College Campus, *Journal of Applied Communication Research*, 37 (4): 465–85.

Carrino, E. A., Bryen, C. P., Maheux, A. J., Stewart, J. L., Roberts, S. R., Widman, L. and Choukas-Bradley, S. (2022), Are Feminists Empowered Activists or Entitled Whiners? A Thematic Analysis of US Adolescents' Definitions of 'Feminist' in a Diverse, Mixed-Gender Sample, *Sex Roles*, 86 (7): 395–414.

Cavalieri, S. (2019), On Amplification: Extralegal Acts of Feminist Resistance in the #MeToo Era, *Wisconsin Law Review*, 1489.

Chirwa, W. C. (2001), Dancing Towards Dictatorship: Political Songs and Popular Culture in Malawi, *Nordic Journal of African Studies*, 10 (1): 1–27.
Chisala, S. (2019), 'Insulting the Modesty of a Woman' and the Push by the Women's Movement in Malawi, *African Feminism*, 9 May.
Chiweshe, M. K. and Bhatasara, S. (2013), Ndezve Varume Izvi: Hegemonic Masculinities and Misogyny in Popular Music in Zimbabwe, *Africa Media Review*, 21 (1): 151–70.
Chunga, J. J. (2020), Malawians Support 2019 Post-Election Demonstrations but Split on Government Power to Limit Protests, *Afrobarometer*, 19 April.
Clark-Parsons, R. (2021), 'I See You, I Believe You, I Stand With You': # MeToo and the Performance of Networked Feminist Visibility, *Feminist Media Studies*, 21 (3): 362–80.
Dlakavu, S. (2016), BLACK FEMINIST REVOLT and Digital Activism Working to End Rape Culture in South Africa, *BUWA! A Journal on African Women's Experiences*, 7 (7): 101–7.
Edwards, A. (1987), Male Violence in Feminist Theory: An Analysis of the Changing Conceptions of Sex/Gender Violence and Male Dominance, in *Women, Violence and Social Control* (pp. 13–29), London: Palgrave Macmillan UK.
Fotopoulou, A. (2016), *Feminist Activism and Digital Networks: Between Empowerment and Vulnerability* (pp. 38–42), London: Palgrave Macmillan.
Fraser, C. (2015), From Ladies First to Asking for it: Benevolent Sexism in the Maintenance of Rape Culture, California Law Review, 103, 141.
Garvey, N. (2005), *Just Sex? The Cultural Scaffolding of Rape*, London/New York: Routledge.
Gaskin, L. (2019), *Rape Culture: Power, Profit, Punishment*, Washington State University.
Kambandu, M. and Manduwi, N. (2022), Are Women Struggling to Break into the Digital in Malawi?, New York: UN Capital Development Fund.
Kember, S. (2002), Reinventing Cyberfeminism: Cyberfeminism and the New Biology, *Economy and Society*, 31 (4): 626–41.
Kemp, S. (2024), Digital 2024: Malawi. https://datareportal.com/reports/digital-2024-malawi.
Kennedy, B. (2000), Cyberfeminisms: Introduction, in *The Cybercultures Reader*, 283–90.
Klason, L. (2016), Connecting Young Women in Malawi to ICTs – Strengthening SRH Information as a Pathway to Empowerment? Lund University.
Kuah-Pearce, K. E. (2008), *Chinese Women and the Cyberspace* (p. 276), Amsterdam University Press.
Malanga, D. F. (2021), Survey of Cyber Violence Against Women in Malawi, *arXiv preprint arXiv:2108.09806*, 22 August.
Malera, G. and Chisala-Tempelhoff, S. (2018), Feminist Contextual Socio-Economic and Political Analysis for Malawi, *JASS*, 30 September. https://justassociates.org/all-resources/feminist-contextual-socio-economic-and-political-analysis-for-malawi/ (Accessed 24 June 2024).
Masina, L. (2018), 'Malawi Government Bans Local Hip-Hop Song for Offending Women, Pan African Visions', *Pan African Visions*, 11 January.
Mawerenga, J. H. (2018), *The Homosexuality Debate in Malawi*, Mzuzu: Mzuni Press.
Mendes, K., Ringrose, J. and Keller, J. (2019), *Digital Feminist Activism: Girls and Women Fight Back Against Rape Culture*, New York: Oxford University Press.
Najumi, M. (2013), We Live in a Rape Culture, *The Feminist Wire*, 9 June.
Ndasauka, Y. (2021), 'Dynamic View of Technology: Implications on Ethics of Social Networking Sites', in Kayange, G. M. and Verharen, C. (eds), *Ethics in Malawi*, 216–29. Washington, DC: The Council for Research in Values and Philosophy.

Ngubane-Mokiwa, S. and Chisale, S. S. (2019), Male Rape or Consensual Sex: Hidden Hegemonic Masculinities by Zulu Speaking Men with Disabilities, *Ubuntu: Journal of Conflict and Social Transformation*, 8 (2): 107–24.

Mkhize, S., Khanyisile, B. M. and Olofinbiyi, S. A. (2022), A Meta-Analytical Study on Rape Culture: Understanding the Myths and the Problems Within the South African Context, *Crime, Criminality and Crime, Criminality and Criminal Justice System*, 49.

Ogbonna, E. C. (2018), Digital Citizenship in Africa's Fractured Social Order, in *Africa's Big Men* (pp. 81–103), London and New York: Routledge.

Pasula, P. (2018), 'Malawi Rights Activists Condemn Mwiza Chavula's Rape Song', *Nyasa Times*, 11 January.

Powell, A. (2015), Seeking Rape Justice: Formal and Informal Responses to Sexual Violence Through Techno Social Counter-Publics, *Theoretical Criminology*, 19 (4): 571–88.

Ringrose, J., Mendes, K., Whitehead, S. and Jenkinson, A. (2021), Resisting Rape Culture Online and At School: The Pedagogy of Digital Defence and Feminist Activism Lessons, in *Violence, Victimisation and Young People: Education and Safe Learning Environments* (pp. 129–153), Cham: Springer International Publishing.

Roberts, T. and Bosch, T. eds (2023), *Digital Citizenship in Africa: Technologies of Agency and Repression*, London: Zed Books.

Rodino-Colocino, M. (2018), Me Too, #MeToo: Countering Cruelty with Empathy, *Communication and Critical/Cultural Studies*, 15 (1): 96–100.

Ryland, M. (2018), *Learning Digital Citizenship in Publics of Practice: How Adults Learn to Use Activist Hashtags on Twitter* (Doctoral dissertation, University of British Columbia).

Sande, N., and Chirongoma, S. (2021), Construction of Rape Culture Amongst the Shona Indigenous Religion and Culture: Perspectives from African Feminist Cultural Hermeneutics, *HTS Teologiese Studies/Theological Studies*, 77 (2).

Semu, L. (2002), Kamuzu's Mbumba: Malawi Women's Embeddedness to Culture in the Face of International Political Pressure and Internal Legal Change, *Africa Today*, 49 (2): 77–99.

Sheik, A. (2018), The More than Beautiful Woman-African Folktales of Female Agency and Emancipation, *Agenda*, 32 (4): 45–53.

Tembo, N. M. (2024), Women, Political Violence, and the Production of Fear in Malawian Social Media Texts, *International Feminist Journal of Politics*, 26 (1): 81–99.

Whaley, R. B. (2001), The Paradoxical Relationship Between Gender Inequality and Rape: Toward a Refined Theory, *Gender & Society*, 15 (4): 531–55.

Yun, J. Y. (2020), Feminist Net-Activism as a New Type of Actor-Network that Creates Feminist Citizenship, *Asian Women*, 36 (4).

Chapter 10

CONTESTING BOUNDARIES: GENDERED CITIZENSHIP AND DIGITAL ACTIVISM IN SUDAN

Maha Bashri

In the midst of Sudan's volatile political landscape, where conflict and patriarchal structures continue to shape public life, Sudanese women are redefining their role in civic and political spaces. While historically constrained by class, gender norms and state repression, Sudanese feminist movements are increasingly leveraging digital platforms to claim visibility, mobilize support and contest exclusionary structures of citizenship. The internet, despite its limitations in Sudan, has become a battleground where gendered citizenship is being renegotiated in real time, offering new avenues for activism and engagement.

This chapter examines the emergence of feminist digital citizenship in Sudan, a concept that encapsulates the ways Sudanese women use digital platforms to assert their rights, build solidarity and challenge patriarchal exclusions (Kadoda and Hale, 2020; Bajec, 2021). Amid ongoing conflict, digital activism is not merely an alternative mode of engagement but a critical survival strategy for women whose political and social agency has long been stifled. This study situates Sudan within broader debates on feminist digital citizenship (Keller, 2012; Fotopoulou, 2016; Jouët, 2018), expanding the discourse beyond Western contexts to explore how digital spaces enable *or* constrain feminist mobilization in conflict-affected societies.

A key question that guides this analysis is whether digital activism in Sudan signals the emergence of a new form of cultural citizenship – one shaped by gendered experiences of conflict, exclusion and resistance (Hopes and Action Foundation Report, 2023). Sudanese women, despite deep-rooted socio-political marginalization, have been at the forefront of digital resistance, leveraging platforms to demand representation, document human rights violations and forge transnational feminist alliances. This chapter interrogates how these digital interventions are shaping feminist digital citizenship and assesses whether they disrupt or reinforce existing socio-economic disparities in Sudan.

To ground this inquiry in concrete cases, the chapter focuses on two prominent organizations: Shabaka and Andariya. These platforms provide distinct yet complementary perspectives on digital feminist activism. Shabaka, with its emphasis on diaspora engagement, facilitates transnational advocacy and crisis response, while Andariya, as a bilingual digital media platform, amplifies cultural and social issues affecting Sudanese women. Through an in-depth analysis of their digital strategies, content creation and engagement practices, this study assesses whether their work fosters a more inclusive expression of Sudanese feminism or inadvertently reproduces entrenched inequalities within digital spaces.

Beyond documenting the efforts of these organizations, this chapter critically examines the structural limitations of digital activism in Sudan. Digital platforms offer opportunities for visibility and mobilization, yet they remain embedded within socio-economic, political and infrastructural constraints. The intersection of class, gender and digital access raises critical questions about who benefits from digital activism and whose voices are amplified or silenced in these online spaces.

The chapter unfolds as follows: First, it provides a historical overview of feminist activism in Sudan, tracing its evolution from grassroots movements to digital engagement. Next, it explores the concept of gendered citizenship in Sudan, analysing how the country's conflict-ridden landscape shapes women's digital participation. The discussion then shifts to the mediation opportunity structure (Cammaerts, 2012) as a framework for understanding Sudanese feminist digital activism, highlighting how women's online mobilization operates within media, discursive and networked opportunity structures. By integrating Cammaerts's mediation theory with Hammett's (2014) conceptualization of active citizenship, this chapter examines how digital feminist activism functions as both a form of protest and a mode of reclaiming citizenship rights.

Ultimately, this chapter contributes to ongoing discussions on feminist digital citizenship, gendered activism and digital rights in conflict settings. It interrogates whether digital platforms are truly expanding the boundaries of women's citizenship and participation in Sudan or whether they remain constrained by historical and structural inequalities. Through the case studies of Shabaka and Andariya, this analysis offers a critical perspective on the transformative possibilities – and limitations – of feminist digital activism in one of Africa's most politically turbulent landscapes.

Feminism disrupted: A historical overview of Sudan's feminist movement

The historical and political context of Sudan has deeply influenced the concept of citizenship, marked by tribal, geographical and gender-based divisions. Prior to British colonization in 1899, the societal structure of Sudan was predominantly shaped by tribal affiliations and geographical ties, playing a crucial role in determining an individual's societal status and belonging. While privileges had always been gendered, favouring males, the colonial administration further complicated this dynamic by selectively granting additional privileges to men from specific tribes and groups. This approach notably favoured the Khatmiyya and Ansar religious sects and the riverine tribes of Northern Sudan (Deng, 1995), thus creating a system of privilege that was both gendered and tribal. These colonial policies laid the groundwork for an exclusionary model of citizenship that systematically marginalized various groups within Sudanese society by gender, tribe and class, thereby reinforcing a patriarchal governance framework that has persisted into contemporary times.

The patriarchal underpinnings of Sudanese citizenship are further evident in the legal sphere, particularly in the debates surrounding family laws (Ibrahim, 2008). Religious communities wield significant influence over the formulation and application of these laws, creating a legal environment that overwhelmingly favours men. The Personal Status Law of 1991, which is based on Islamic Sharia law, epitomizes this gender disparity by granting men the rights to polygamy and unilateral divorce while imposing stringent restrictions on women's personal freedoms (Nageeb, 2004; Tønnessen, 2011; Tønnessen and Kjøstvedt, 2010). Such legal disparities underscore the unequal status of women within the citizenship framework and highlight the role of religious dictates in shaping the contours of citizenship in Sudan.

Sudan's feminist movement has its roots in the early twentieth century, driven by the intellectual and strategic efforts of a well-educated elite (Ahmed, 2014; Seesemann, 2005). Initially intertwined with nationalist agendas that did not explicitly prioritize women's rights, the movement began to take a more defined shape around 1945, with the emergence of the first women's organizations. These groups, pioneered by influential female leaders, such as the late Fatima Ibrahim Ahmed, and often led by trade unions and the Sudanese Communist Party, laid the foundational groundwork for structured feminist activism (Hale, 2004). The Sudanese Women's Union, established in 1952, marked a pivotal moment in the movement's history, focusing initially on issues of health, education, equal pay and female literacy. By the 1953 elections, the Sudanese Women's Union had expanded its advocacy to include women's voting rights, three years before Sudan's independence (Abbas, 2010). Despite these advancements, the feminist movement in Sudan has been criticized for reflecting predominantly the perspectives of middle-class intellectuals rather than the daily experiences of the wider populace (Bayat, 2000).

During Omer al-Bashir's regime following the 1989 coup by the National Islamist Front, the feminist movement encountered renewed challenges. The regime's Islamization project imposed restrictive interpretations of Sharia law that significantly curtailed women's rights. In response, women adapted by leveraging emerging digital platforms and grassroots movements, such as the No to Women's Oppression Initiative, which became central to the feminist struggle from the 1990s until the December revolution in 2018 (Bashri, 2023; Skalli, 2014; Coslovi et al., 2020).

The December 2018 revolution represented a significant shift, as women traditionally excluded from political participation were now at the forefront of activism. The Sudanese feminist movement, increasingly inclusive of women from diverse backgrounds, classes and regions beyond Khartoum post-Bashir, faced ongoing challenges in securing leadership roles within the evolving political landscape despite the transformative potential of digital activism, which facilitated broader connectivity and mobilization (Reliefweb, 2023).

In the December revolution in Sudan, women played pivotal roles that were as diverse as they were influential (Malik, 2019). Young women across the country broke away from traditional gender roles, stepping into leadership positions, organizing logistics and leading digital media campaigns. This active participation challenged the established norms and illustrated a crucial shift in the public perception of women's capabilities and rights. Their involvement not only highlighted their agency but also underscored a shift towards a more active form of citizenship, where women are seen not just as passive recipients of state policies but as essential, dynamic agents of change (Abdalla et al., 2023).

Despite these advances, the transitional period post-revolution revealed persistent challenges. Women, particularly young and active participants in the revolution, found themselves sidelined in political decision-making processes that remained male-dominated. This marginalization was felt despite their significant contributions and highlighted the ongoing struggle against a patriarchal framework that continues to underestimate and undervalue women's roles in public and political spheres. The exclusion of women from formal politics, justified often by their supposed lack of experience or the conservative nature of Sudanese culture, accentuated the deep-rooted patriarchal barriers still present in Sudan (Kadoda and Hale, 2020).

The revolution also catalysed a generational shift within the feminist movement in Sudan. Younger women activists began challenging both the traditional societal norms and the older generation's more conservative feminist approaches. This shift is indicative of a broader, more inclusive and assertive feminist mobilization, striving not only for gender equality but also for a re-evaluation of what feminist activism encompasses. The movement's inclusivity is demonstrated by its representation of women from different social classes and geographical areas throughout the country, ensuring a more comprehensive approach to feminist issues in Sudan. This new wave of feminism is marked by a willingness to confront and transform the patriarchal structures that have long governed Sudanese society (Nugdalla, 2020).

Regionally and culturally, the participation of women in the revolution varied significantly, reflecting Sudan's diverse cultural landscape. In some areas, merely participating in protests was a radical act, challenging deep-seated cultural norms about women's visibility and roles in public spaces. This regional diversity highlights the complexity of advocating for gendered citizenship in Sudan and emphasizes the need for nuanced approaches that consider the cultural and social specifics of different regions. In this context, gendered citizenship refers to a state where women have the same rights as their male counterparts while taking into account the unique social and cultural contexts within Sudan. This approach recognizes that achieving equal rights and opportunities for women must be tailored to the diverse cultural landscapes and social norms across different parts of the country, ensuring that the pursuit of gender equality is both effective and culturally sensitive.

Post-revolution, there was a noticeable surge in the formation of new feminist groups that not only focused on immediate practical needs but also aimed to address strategic interests that seek to transform the underlying gender relations. These groups advocated for a fundamental change in how women's roles and rights are perceived and implemented in Sudan, aiming for a lasting impact on the societal and political structures. The emergence of intersectional feminist approaches (Crenshaw, 1989) within these new groups points to what was considered a promising direction in Sudan's feminist movements (pre-conflict). By acknowledging and addressing the diverse experiences of women across different social, economic and ethnic backgrounds, these groups paved the way for a more inclusive and effective feminist activism that not only challenged existing gender disparities but also integrated broader social justice issues into the feminist agenda.

Analysing the trajectory of the Sudanese feminist movement provides a critical lens through which to explore and understand the intricate dynamics of citizenship in the Sudanese context. Historically, the movement has branched into two main paths, each characterized by distinct ideological underpinnings but converging in their predominantly middle-class, elite composition (Bashri, 2023; Tønnessen, 2011). Islamist women within the movement have concentrated their efforts on securing gender equity within the family unit, pushing for reforms that are coherent with Islamic tenets. In contrast, secular factions have aimed to extend the scope of their advocacy to achieve societal gender equality that transcends familial boundaries and impacts broader public and political spheres.

Despite the distinct goals and methods of these two strands, both grapple with the common challenge of navigating a societal framework heavily influenced by traditional and religious structures. This convergence underlines a shared struggle across the feminist spectrum, illustrating how both Islamist and secular groups, despite their differences, are rooted in the perspectives of Sudan's middle-class elite, predominantly concentrated in the centre of Khartoum, the capital. This centralization often excludes voices from other regions and ethnic groups, particularly those from marginalized areas outside the capital.

Such dynamics are central to understanding the multifaceted nature of the feminist movement in Sudan, where the intersections of gender, religion, class and societal norms are pivotal in shaping their quest for equality and justice (Layish and Warburg, 2002; Narayan, 1997, 2004).

This examination segues into a broader discussion of citizenship in Sudan, which has historically been defined by the state's distribution of legal rights rather than by active participation from its citizens. Citizenship, as conceptualized by Kymlicka and Norman (1994), is both a legal status and an identity reflecting one's membership in a community, embodying principles of inclusion and exclusion. In Sudan, women, like many of their counterparts in the Global South, frequently find themselves categorized as 'excluded', relegated to what can be described as passive citizenship. They receive certain protections and benefits from the state yet are largely marginalized from actively engaging in public discourse and shaping state policies (Bhargava, 2005; Jayal, 2001). This passive form of citizenship not only highlights the historical and ongoing barriers Sudanese women face in asserting their roles and rights but also underscores the broader implications of how citizenship and rights are distributed – and *perhaps* contested and claimed – within the Sudanese political landscape.

Gendered citizenship in Sudan's nation state: A critical examination

The concept of citizenship is often framed as a universal, legal and political status, yet in practice, it is shaped by intersecting power structures that dictate who is included or excluded from full civic participation. In Sudan, citizenship has historically been constructed through a gendered and exclusionary framework, privileging men's participation in public life while relegating women to passive recipients of state policies. These exclusions are not merely legal but are deeply embedded in social, cultural and economic structures, reinforcing women's marginalization in the public sphere (Seidman, 1999; Orloff, 1993).

Traditional citizenship models in Sudan reflect what Kymlicka and Norman (1994) describe as a state-centric, legalistic approach that grants formal rights while ignoring the structural inequalities that limit certain groups' ability to exercise these rights. Feminist scholars have long critiqued such models, arguing that citizenship is not neutral but gendered, as the state often enforces policies that privilege men's participation while constraining women's agency (Pateman, 1988; Young, 1993).

In Sudan, gendered citizenship manifests through legal, political and economic mechanisms that systematically exclude women from decision-making and public life. The Personal Status Law of 1991, which remains in effect, exemplifies these exclusions by codifying men's dominance over marriage, divorce and family structures (Tønnessen and Kjøstvedt, 2010). This legal framework reinforces patriarchal control over women's autonomy, limiting their ability to engage as full citizens.

At the same time, Sudanese citizenship has historically been defined by ethnic, tribal and class-based divisions, which intersect with gender to create multilayered exclusions (Deng, 1995). These intersections highlight what Jayal (2001) terms 'differentiated citizenship', where legal rights exist but are unequally distributed based on social positioning. For Sudanese women, particularly those from marginalized communities outside Khartoum, these layers of exclusion reinforce their status as subordinate citizens.

Women's citizenship in Sudan has largely been characterized by passivity, where they are recognized as legal subjects but denied full participation in political and economic decision-making (Bhargava, 2005). As Nyamnjoh (2007) argues, traditional citizenship models in many postcolonial African states are 'bounded', restricting women's participation by relegating them to private or domestic roles. Sudan's legal and political institutions reinforce this passivity through policies that limit women's access to land ownership, inheritance rights and political office (Tønnessen, 2011).

However, feminist scholars argue that citizenship must be reconceptualized beyond legal status to include the everyday practices through which marginalized groups claim rights and negotiate their place in society (Isin, 2009; Hammett, 2014). This shift from passive to active citizenship is particularly crucial in Sudan, where women have historically challenged patriarchal exclusions through grassroots mobilization, protest movements and digital activism.

The December 2018 revolution demonstrated women's capacity for active citizenship, as they took leading roles in protests, digital mobilization and political organizing (Abdalla et al., 2023). However, the post-revolution period revealed the fragility of these gains, as women were largely sidelined in transitional governance structures despite their activism (Kadoda and Hale, 2020). This exclusion underscores a fundamental tension: women are often central to political change but remain marginalized in the formal distribution of power.

The persistence of legal, cultural and economic barriers means that Sudanese women's citizenship remains precarious and conditional, shaped by state policies, social norms and ongoing conflict. Women continue to navigate a political landscape that acknowledges their presence but restricts their agency, a phenomenon that aligns with Narayan's (1997, 2004) critique of postcolonial citizenship – where women are visible in national narratives but excluded from substantive political participation.

At the same time, Sudanese women have resisted these exclusions by leveraging digital platforms to assert their political agency, document rights violations and build transnational feminist networks (Bajec, 2021; Hopes and Actions Foundation, 2023). This shift raises key questions about whether digital spaces are creating a new form of feminist digital citizenship that disrupts traditional state-controlled definitions of belonging and rights.

To move beyond Sudan's exclusionary citizenship model, feminist scholars emphasize the need to redefine citizenship as an ongoing, negotiated process that incorporates diverse gendered experiences (Chari, 2009; Seidman, 1999). This requires moving away from a legalistic framework and instead recognizing how women actively shape citizenship through protest, digital activism and community organizing.

Nyamnjoh's (2007) concept of flexible citizenship is useful in this context, as it suggests that citizenship is not static but continually contested and reshaped. Sudanese women, particularly through digital activism, are exemplifying this flexibility, using online spaces to redefine their political participation, challenge patriarchal state structures and mobilize across borders (Keller, 2012; Cammaerts, 2012).

However, the question remains whether digital activism alone can dismantle the deeply entrenched gendered exclusions within Sudan's citizenship framework. While online activism expands visibility and discourse, its ability to translate into formal legal and policy changes remains uncertain, particularly in a state where political instability and military dominance continue to shape governance (Tønnessen, 2011).

Sudanese women's citizenship has long been constrained by legal, economic and political exclusions, yet they continue to contest these restrictions through activism, both offline and online. The 2018 revolution and subsequent digital feminist movements reveal a push towards a more inclusive, participatory citizenship, but structural barriers persist. The future of gendered citizenship in Sudan depends on whether feminist activism – both in the streets and in digital spaces – can translate symbolic resistance into substantive legal and political change.

Navigating digital resistance: Feminist activism and citizenship in Sudan

Feminist digital activism in Sudan operates within a highly contested media and political landscape, where activists must navigate state repression, patriarchal norms and digital inequalities. Women's access to civic participation has long been constrained by restrictive laws, cultural expectations and exclusion from formal political spaces. However, the digital sphere has emerged as an alternative battleground, where Sudanese feminist activists are redefining citizenship, mobilizing networks and contesting exclusion.

To analyse these dynamics, this section applies Cammaerts's (2012) mediation opportunity structure framework, which provides a lens for understanding how activists use media, discourse and networks to advance their causes. Cammaerts identifies three key dimensions: the media opportunity structure, which refers to activists' ability to access and influence media narratives; the discursive opportunity structure, which captures the production of counter-narratives that challenge dominant ideologies; and the networked opportunity structure, which focuses on how activists organize, mobilize and sustain networks through digital platforms. By applying this framework to Sudanese feminist activism, it is possible to examine how women leverage digital tools not only to protest gendered injustices but also to redefine what it means to be a citizen in an exclusionary state.

Keller (2012) defines feminist digital citizenship as the use of digital technologies to claim rights, participate in civic life and challenge gender inequalities. This concept bridges Cammaerts's framework with broader theories of citizenship. For instance, Rosaldo's (1994) concept of cultural citizenship emphasizes how cultural practices shape belonging and participation. In the Sudanese context, digital platforms become spaces where women assert their rights and belonging within the national community, effectively leveraging the discursive opportunity structure identified by Cammaerts.

Isin's (2009) concept of activist citizenship further enriches this understanding by highlighting practices that challenge established citizenship norms and create new political spaces. This aligns closely with how Sudanese feminist activists use digital platforms to contest discriminatory laws, discuss women's political representation and document women's contributions to social movements. These actions exemplify how activists utilize the networked opportunity structure to enact activist citizenship.

Bennett and Segerberg's (2012) theory of digital mobilization provides insight into how Sudanese feminist activists organize and coordinate collective action despite physical and social barriers. This concept illuminates the practical application of Cammaerts's networked opportunity structure, showing how digital tools enable new forms of organization and participation that are more flexible and less hierarchical than traditional social movements.

Hammett's (2014) exploration of active and activist citizenship complements Isin's work and provides a bridge between Cammaerts's framework and citizenship theories. Hammett's distinction between state-promoted active citizenship and grassroots activist citizenship helps explain how Sudanese feminists use digital spaces to contest official narratives and engage in political participation beyond conventional boundaries. This approach demonstrates how activists leverage all three components of the mediation opportunity structure to enact feminist digital citizenship.

The affordances of digital technologies that enable new forms of feminist digital citizenship in Sudan – such as anonymity, rapid information sharing, low-cost advocacy and digital documentation – can be understood as practical manifestations of the opportunities presented within Cammaerts's framework. These affordances allow Sudanese feminist activists to navigate and exploit media, discursive and networked opportunity structures effectively.

By integrating these theoretical perspectives, it becomes clear that Sudanese feminist digital activists navigate and leverage media, discursive and networked opportunity structures to claim cultural and activist citizenship. This synthesis provides a robust framework for analysing how feminist digital activism is reshaping notions of citizenship and political participation in Sudan. It also demonstrates that the digital public sphere is not just a platform for activism but a space where new forms of citizenship are being negotiated and enacted through the strategic use of media and communication opportunities.

Sudan's digital landscape: Connectivity and activism

To contextualize my analysis, it is crucial to understand Sudan's digital landscape. Prior to the conflict that erupted in April 2023, internet penetration in Sudan was already low compared to global standards. According to DataReportal (2023), as of January 2023, there were approximately 13.49 million internet users in Sudan, representing a penetration rate of about 28.4 per cent of the total population. This left nearly 71.6 per cent without internet access at the start of the year, with only modest growth in user numbers from 2022 to 2023.

The onset of conflict in April 2023 further impacted internet connectivity across Sudan. Both the Sudanese Armed Forces and the Rapid Support Forces reportedly employed internet shutdowns as a strategy to control information flow. These shutdowns are part of a broader pattern in Sudan, with at least ten major internet disruptions recorded between 2016 and 2021 (Mohamed, 2021). Such disruptions severely limit activists' ability to organize and communicate, directly impacting the networked opportunity structure.

Even before the conflict, there was a notable gender disparity in internet access, with a 10 per cent difference in regular internet usage between males and females (Lardies, Dryding and Logan, 2019; Majama, 2019). This disparity likely widens under conflict conditions, as economic pressures and security concerns disproportionately affect women's access to digital resources.

Despite these challenges, digital activism has played a significant role in Sudan's recent political landscape. The 2018–2019 Sudanese revolution saw widespread use of social media for organizing protests, sharing information and documenting human rights abuses (Mahmoud, 2021). Women played a crucial role in this digital activism, using platforms like Facebook and Twitter to mobilize support and challenge gender norms (Kadoda and Hale, 2020).

In the current conflict, Sudanese women's digital activism has taken various forms, including the following:

1. Social media campaigns raising awareness about human rights violations
2. Online fundraising efforts for humanitarian aid
3. Digital documentation of women's experiences during the conflict
4. Virtual forums for discussing and shaping post-conflict visions for Sudan

The role of Sudanese women in the diaspora, who often retain internet access, has become increasingly crucial. Their digital activism represents a powerful form of resistance against the ongoing conflict and the disproportionate victimization of women. However, this also raises questions about representation and the potential 'weaponization' of privilege by the elite middle class in their efforts to advocate for all Sudanese women.

By examining these dynamics, I aim to contribute to a deeper understanding of feminist digital citizenship in Sudan's conflict and limited digital access. The following research questions will guide this investigation:

RQ1: How are Sudanese women leveraging digital platforms to articulate gendered experiences and challenge existing power structures amid the ongoing conflict?
RQ2: To what extent is a gendered solidarity emerging through digital activism in the context of the current crisis?
RQ3: How are privileged Sudanese women, particularly those in the diaspora, using their access to digital platforms to advocate for broader women's rights and representation?
RQ4: What potential do these digital activism practices hold for redefining citizenship rights in a future, potentially peaceful Sudan?

Methodology

This study employed a qualitative multi-method approach to examine the digital activism of Andariya and Shabaka, focusing on document analysis, social media content analysis and semi-structured interviews. A review of each organization's background, mission and digital presence provided context for their advocacy efforts. Social media and website content were analysed to assess how the organizations use digital platforms to advance feminist causes. Semi-structured interviews with key figures offered insights into their strategies and challenges, as well as the impact of ongoing conflict on women's rights. Thematic analysis identified patterns in their digital activism, highlighting common strategies and distinct challenges in promoting gender equality and reshaping gendered citizenship in Sudan.

Analysis

The analysis of Andariya and Shabaka highlights how Sudanese feminist organizations use digital platforms to redefine citizenship and challenge patriarchal norms amid ongoing conflict. Cammaerts's (2012) mediation opportunity structure framework, alongside feminist digital citizenship (Keller, 2012) and activist citizenship (Isin, 2009; Hammett, 2014), provides a lens to examine how these organizations navigate and reshape media, discursive and networked opportunities within Sudan's constrained digital landscape.

Media opportunity structure

Andariya's bilingual digital platform embodies feminist digital citizenship by allowing marginalized voices to challenge dominant narratives through digital media (Keller, 2012). Given Sudan's history of excluding women from public discourse, this work is particularly significant (Bashri, 2023; Kadoda and Hale, 2020). Publishing in both Arabic and English ensures accessibility for local audiences while fostering transnational feminist solidarity.

During the December revolution, Andariya's documentation of women's participation challenged traditional media portrayals. This period saw unprecedented recognition of women's political agency, particularly through coverage of neighbourhood resistance committees where women played leading roles (Kadoda and Hale, 2020). Hale (2004) previously analysed Sudan's tendency to minimize women's political contributions, a pattern Andariya actively disrupts.

Shabaka, in contrast, centres its media strategy on transnational advocacy and crisis response. The Sudan Crisis Coordination Unit reflects mediated citizenship, which Hammett (2014) describes as activism emerging within digital and political constraints. Similarly, political transitions often hinge on mediated forms of activism (O'Donnell and Schmitter, 1986). Shabaka's digital work during the conflict has been instrumental in sustaining feminist activism when traditional political spaces are disrupted (Reliefweb, 2023).

The structural challenges facing both organizations have become even more pronounced as Sudan's war enters its second year. Internet shutdowns, long used as a tool of political control, have escalated during the conflict, severely disrupting digital activism at a time when reliable communication is most critical (Access Now, 2024). Between 2016 and 2021, at least ten major disruptions were documented (Mohamed, 2021), but since the war began, restrictions have become more frequent and prolonged, further isolating activists and limiting their ability to mobilize. These shutdowns exemplify the repressive structure that constrains media opportunities (Cammaerts, 2012), with devastating consequences for feminist and civil society organizations trying to document atrocities, coordinate aid and advocate for political change. Sudan's already low internet penetration rate of 28.4 per cent (DataReportal, 2023) compounds the problem, with urban–rural and gender divides further restricting access, leaving many women – especially in conflict-affected areas – without digital lifelines.

Discursive opportunity structure

The strategies employed by Andariya and Shabaka reflect the complex relationship between digital activism and citizenship rights. Andariya's content aligns with the concept of networked feminist publics, where digital spaces facilitate alternative narratives of Sudanese womanhood (Fotopoulou, 2016). This work directly challenges the historically masculinized conception of citizenship, particularly in the context of Sudan's Personal Status Law of 1991 (Tønnessen, 2011). The platform's bilingual cultural production also reflects broader feminist movements in North Africa, where local–global connections are crucial (Skalli, 2014). By documenting the cultural and social realities of Sudanese women, Andariya fosters cultural citizenship, enabling marginalized groups to assert their identities while demanding full political rights (Rosaldo, 1994).

Meanwhile, Shabaka amplifies diaspora voices while maintaining connections to on-the-ground realities in Sudan. This model aligns with flexible citizenship, where transnational actors shape local political activism (Nyamnjoh, 2007). The role of diaspora networks in sustaining feminist activism is particularly crucial during political repression (Coslovi et al., 2020). Historical divisions within Sudan's feminist movement – especially between secular and Islamist perspectives – further complicate digital advocacy (Bashri, 2023).

However, Sudan's digital divide influences how these discursive strategies function. A documented 10 per cent gender gap in internet usage limits women's participation (Lardies, Dryding and Logan, 2019). Broader technological marginalization across Africa affects women's ability to engage in digital activism (Majama, 2019). These inequalities reflect differentiated citizenship, where social and economic hierarchies determine access to political expression (Jayal, 2001).

Networked opportunity structure

The ability to leverage digital networks has been crucial during Sudan's ongoing conflict, though it remains constrained by technological and social barriers.

Shabaka's Crisis Coordination Unit demonstrates connective action, which allows for rapid mobilization and cross-border coordination (Bennett and Segerberg, 2012). In situations where traditional organizing is disrupted, digital networks have become central to sustaining feminist activism (Abdalla et al., 2023).

Shabaka's work also illustrates acts of citizenship, where individuals or groups claim new rights and responsibilities through digital activism (Isin, 2009). Chari (2009) highlights that gendered citizenship requires both political claims-making and solidarity-building, both of which Shabaka's digital initiatives facilitate. Andariya, by contrast, focuses on bridging urban and rural experiences, addressing the intersections of gender, class and geography in shaping citizenship (Seidman, 1999). Yet, significant access barriers remain. Women in conflict-affected areas face particularly acute limitations, as documented by the Hopes and Actions Foundation (2023). The exclusion of these groups raises concerns about the reach and inclusivity of digital feminist networks.

Rethinking feminist digital citizenship in Sudan

The analysis highlights key tensions in feminist digital citizenship in Sudan, where digital platforms provide new opportunities for feminist engagement but also risk reinforcing existing inequalities. Even when legal rights are formally recognized, they do not always translate into meaningful political power, a challenge well documented in discussions of gendered citizenship (Tønnessen, 2011). Digital activism, while a powerful tool for visibility and mobilization, is constrained by class-based disparities in access to technology (Bashri, 2023), raising questions about whose voices are amplified and whose struggles remain invisible.

Representation remains a central issue, particularly in the context of postcolonial feminist activism. Spivak's (1988) concerns about the politics of representation resonate in Sudan, where diaspora voices often shape narratives about Sudanese women's experiences in ways that may not always align with local priorities (Coslovi et al., 2020). This tension between local and transnational feminist agendas is not unique to Sudan but is a broader challenge in postcolonial feminist movements (Narayan, 1997, 2004). While transnational networks can offer crucial solidarity and advocacy platforms, they also risk overshadowing or diluting local feminist struggles.

Sustainability is another critical concern. While digital activism has been instrumental in amplifying Sudanese women's voices, lasting political change requires more than online mobilization – it depends on strong, sustained organizational structures (Tufekci, 2017). Sudanese women have played significant roles in revolutionary movements, yet their participation has not always translated into long-term political influence (Abdalla et al., 2023). This highlights a recurring challenge in digital mobilization: visibility alone is insufficient without institutional and structural change to sustain feminist gains.

This chapter has examined how Sudanese women leverage digital platforms to articulate gendered experiences and challenge patriarchal power, particularly in the context of conflict. While digital activism creates new forms of feminist citizenship that transcend traditional boundaries, its effectiveness is deeply shaped by broader social and political realities. Kadoda and Hale's (2020) work on Sudanese women's 'radical imaginations' underscores how digital activism both empowers and limits feminist mobilization, offering new spaces for political participation while also reflecting the constraints of the existing socio-political order.

These findings contribute to ongoing discussions on feminist digital citizenship in conflict settings, emphasizing the need to address both technological and social barriers to women's political participation. While digital platforms can serve as sites of resistance and solidarity, they also mirror and sometimes reinforce existing social hierarchies (Tønnessen, 2011). To truly transform citizenship practices, feminist digital activism must not only navigate repressive political structures but also ensure inclusivity across class, geography and access to technology.

Future research should examine how digital feminist initiatives translate into sustained political change, particularly in post-conflict settings. This requires exploring the intersections between feminist digital citizenship and broader political transformations (O'Donnell and Schmitter, 1986) while remaining attentive to Sudan's specific socio-political context. Understanding these dynamics will be crucial in assessing whether digital activism can move beyond moments of mobilization to create enduring shifts in women's political power and rights.

Bibliography

Abbas, S. (2010), The Sudanese Women's Movement and the Mobilisation for the 2008 Legislative Quota and its Aftermath, *IDS bulletin*, 41 (5): 100–8.

Abdalla, Y. E., Bashir, H. H., Mohammed, I. A. O., Adam, Z. O. M., Al Tirifi, H. I., Babiker, S. Y., Batio, A. A. H., with the contribution of al-Nagar, S., and Tønnessen, L. (2023), *A Regional Insight into Sudanese Women's Participation in the December Revolution* (Sudan Working Paper No. 1). The Sudan-Norway Academic Cooperation Programme. Royal Norwegian Embassy in Khartoum. ISBN 978-82-8062-834-3 (PDF). ISSN 1890-5056.

Access Now (2024), #KeepItOn in Times of War: Sudan's Communications Shutdown Must be Reversed Urgently, April 12. https://www.accessnow.org/press-release/keepiton-sudan-shutdown/.

Ahmed, M. M. E. (2014), The Women's Movement in Sudan From Nationalism to Transnationalism: Prospects for a Solidarity Movement, 4 December.

Bajec, A. (2021), *The Women of Sudan are Pushing for a Feminist Agenda*, Equal Times, July 12. Retrieved March 21, 2024, from https://www.equaltimes.org/the-women-of-sudan-are-pushing-for?lang=en.

Bashri, M. (2023), 'Noon Al Niswa' – N is for the Female Collective: Contesting Androcentric Power Structures Through Grassroots Women's Groups in Sudan, *Information, Communication & Society*, 26 (5): 1–16. doi:10.1080/1369118X.2023.2224418.

Bayat, A. (2000), From Dangerous Classes to Quiet Rebels Politics of the Urban Subaltern in the Global South, *International Sociology*, 15 (3): 533–57.

Bennet, W. L. and Segerberg, A. (2012), The Logic of Connective Action, *Information, Communication & Society*, 15 (5): 739–68.

Bhargava, R. (2005), Introduction, in R. Bhargava and H. Reifeld (eds), *Civil Society, Public Sphere and Citizenship: Dialogues and Perceptions* (1–13), London: Sage Publications.

Cammaerts, B. (2012), Protest Logics and the Mediation Opportunity Structure, *European Journal of Communication*, 27 (2): 117–34.

Chari, A. (2009), Gendered Citizenship and Women's Movement, *Economic and Political Weekly* 44 (17), 47–57.

Coslovi, L., Ianni, A., Marcheggiani, E. and El Kadi, T. H. (2020), *Women in Transition: The Role of Women and the Arab Springs 2.0 in Sudan and Algeria*, Centro Studi di Politica Internazionale (CeSPI), December. https://www.cespi.it/en/projects/women-transition-role-women-and-arab-springs-20-sudan-and-algeria.

Crenshaw, K. (1989), Demarginalizing the Intersection of Race and Sex: A Black Feminist Critique of Antidiscrimination Doctrine, Feminist Theory and Antiracist Politics, *University of Chicago Legal Forum*, 149, 139–67.

DataReportal (2023), *Digital 2023: Sudan*, 14 February. https://datareportal.com/reports/digital-2023-sudan.

Deng, F. M. (1995), *War of Visions: Conflict of Identities in the Sudan*, Washington, DC: Brookings Institution Press.

Fotopoulou, A. (2016), *Feminist Activism and Digital Networks: Between Empowerment and Vulnerability*, London: Palgrave Macmillan.

Gerbaudo, P. (2012), *Tweets and the Streets: Social Media and Contemporary Activism*, London: Pluto Press.

Hale, S. (2004), *Gender Politics in Sudan: Islamism, Socialism, and the State*, London and New York: Routledge.

Hammett, D. (2014), Understanding the Role of Communication in Promoting Active and Activist Citizenship, *Geography Compass*, 8 (9).

Hopes and Actions Foundation (2023), *Breaking Barriers: Examining the Digital Exclusion of Women and Online Gender-Based Violence in Sudan*, Association for Progressive Communications. https://firn.genderit.org/sites/default/files/2023-09/Breaking_barriers_report.pdf.

Ibrahim, A. A. (2008), *Manichaean Delirium: Decolonizing the Judiciary and Islamic Renewal in the Sudan, 1898–1985*, Leiden: Brill.

Isin, E. F. (2009), Citizenship in Flux: The Figure of the Activist Citizen, *Subjectivity*, 29 (1): 367–88.

Jayal, G. N. (2001), Reinventing the State: The Emergence of Alternative Models of Governance in India in the 1990s, in N. G. Jayal and S. Pai (Eds), *Democratic Governance in India: Challenges of Poverty, Development, and Identity* (50–71), London: Sage Publications.

Jouët, J. (2018), Digital Feminism: Questioning the Renewal of Activism, *Journal of Research in Gender Studies*, 8 (1): 133–57.

Kadoda, G. and Hale, S. (2020), The Radical Imaginations of Sudanese Women: A Gendered Revolution, *Al-Raida Journal*, 44 (1): 73–92.

Keller, J. M. (2012), Virtual Feminisms: Girls' Blogging Communities, Feminist Activism, and Participatory Politics, *Information, Communication & Society*, 15 (3): 429–47.

Kymlicka, W. and Norman, W. (1994), Return of the Citizen: A Survey of Recent Work on Citizenship Theory, *Ethics*, 104 (2): 352–81. http://dx.doi.org/10.1086/293605.

Lardies, C. A., Dryding, D. and Logan, C. (2019), *Gains and Gaps: Perceptions and Experiences of Gender in Africa*, Afrobarometer Policy Paper No. 61, Afrobarometer. https://www.afrobarometer.org/wp-content/uploads/migrated/files/publications/Policy%20papers/ab_r7_policypaperno61_gains_and_gaps_gender_perceptions_in_africa.pdf.

Layish, A. and Warburg, G. R. (2002), *The Reinstatement of Islamic Law in Sudan Under Numayrī*, Leiden: Brill.

Malik, N. (2019), She's an Icon of Sudan's Revolution. But the Woman in White Obscures Vital Truths, *The Guardian*, 24 April. https://www.theguardian.com/commentisfree/2019/apr/24/icon-sudan-revolution-woman-in-white.

Majama, K. (2019), African Women Face Widening Technology Gap, *African School on Internet Governance (AfriSIG)*, April 1. https://afrisig.org/2019/04/01/african-women-face-widening-technology-gap/.

Mohamed, A. (2021), Sudan Digital Rights Landscape Report, in T. Roberts (ed.), *Digital Rights in Closing Civic Space: Lessons from Ten African Countries*, 105–24, Brighton: Institute of Development Studies.

Nageeb, S. A. (2004), *New Spaces and Old Frontiers: Women, Social Space, and Islamization in Sudan*, Lanham, MD: Lexington Books.

Narayan, U. (1997), Towards a Feminist Vision of Citizenship: Rethinking the Implications of Dignity, Political Participation and Nationality, in U. Narayan and M. L. Shanley (eds), *Reconstructing Political Theory: Feminist Perspectives*, 48–64, University Park, PA: Pennsylvania State University Press.

Narayan, U. (2004), The Project of Feminist Epistemology: Perspectives from a Nonwestern Feminist, in S. Harding (ed.), *The Feminist Standpoint Theory Reader: Intellectual and Political Controversies*, 213–24, London and New York: Routledge.

Nugdalla, S. O. (2020), The Revolution Continues: Sudanese Women's Activism, in A. Okech (ed.), *Gender, Protests and Political Change in Africa*, 107–30, Cham, CH: Palgrave Macmillan.

Nyamnjoh, F. B. (2007), From Bounded to Flexible Citizenship: Lessons from Africa, *Citizenship Studies*, 11 (1): 73–82.

O'Donnell, G. and Schmitter, P. C. (1986), *Transitions from Authoritarian Rule: Tentative Conclusions about Uncertain Democracies*, Baltimore, MD: Johns Hopkins University Press.

Orloff, A. S. (1993), Gender and the Social Rights of Citizenship: The Comparative Analysis of Gender Relations and Welfare States, *American Sociological Review*, 58 (3): 303–28.

Pateman, C. (1988), *The Sexual Contract*, Cambridge, UK: Polity Press.

Reliefweb. (2023), Women on the Frontlines: A Feminist Perspective on the Ongoing Crisis in Sudan, 26 April. https://reliefweb.int/report/sudan/women-frontlines-feminist-perspective-ongoing-crisis-sudan.

Rosaldo, R. (1994), Cultural Citizenship in San Jose, California, *PoLAR: Political and Legal Anthropology Review*, 17 (2): 57–64.

Seesemann, R. (2005), The Saintly Woman and the Politician: Fatima Talib and the Sudanese Women's Movement, in B. Soares and R. Otayek (eds), *Islam and Muslim Politics in Africa*, 159–80, New York: Palgrave Macmillan.

Seidman, G. W. (1999), Gendered Citizenship: South Africa's Democratic Transition and the Construction of a Gendered State, *Source: Gender and Society*, 13 (3): 287–307.

Sidahmed, A. S. (2020), The Sudanese Women's Movement: Challenges and Prospects, *Ahfad Journal*, 37 (1): 1–18.

Siddiqui, U. and Mohamed, H. (2023), Sudan Updates: Internet Outage Reported Across the Country, *Al Jazeera*, April 23.

Skalli, L. H. (2014), Young Women and Social Media Against Sexual Harassment in North Africa, *The Journal of North African Studies*, 19 (2): 244–58.

Spivak, G. C. (1988), Can the Subaltern Speak?, in C. Nelson and L. Grossberg (eds), *Marxism and the Interpretation of Culture*, 271–313, London: Macmillan Education.

Tønnessen, L. (2011), Beyond Numbers? Women's 25% Parliamentary Quota in Post-Conflict Sudan, *Journal of Peace, Conflict and Development*, 17.

Tønnesen, L. and Kjøstvedt, H. G. (2010), The Politics of Women's Representation in Sudan: Debating Women's Rights in Islam from the Elites to the Grassroots, CMI Report R, 2010: 2.

Tønnessen, L. (2008), Gendered Citizenship in Sudan: Competing Debates on Family Laws Among Northern and Southern Elites in Khartoum, *The Journal of North African Studies*, 13 (4): 455–69.

Tufekci, Z. (2017), *Twitter and Tear Gas: The Power and Fragility of Networked Protest*, New Haven: Yale University Press.

Young, K. (1993), *Planning with Women: Making a World of Difference*, London: Macmillan.

Chapter 11

DIGITAL FEMINISM IN ETHIOPIA

Selamawit Tezera Chaka

In the context of collective digital feminist organizing, Ethiopia has been trailing most African nations. This is primarily attributed to the limited internet penetration and low level of digital literacy in the country. Although data varies, it is estimated that less than 25 per cent of the population has internet access, with the majority of users residing in the city centres. This digital divide hampers effective individual participation in social-media-based activism. Despite these challenges, a few impactful social-media-based feminist movements have emerged in Ethiopia since 2014 (Halefom, 2017; Zewde, 2019).

This paper will explore the utilization of social media as a line for advancing the feminist movement in Ethiopia. It will present a comprehensive argument regarding the impact, learning outcomes and challenges faced by digital feminist movements as a result of their engagement in social media as well as events on the ground. The study will draw attention to the works of various organizations and initiatives, including Yikono, the Yellow Movement, Africanfeminism.com (which used to be under the custodianship of Earuyan Solutions), Addis Powerhouse, Article 35 and Merahit. These organizations have been selected for their significant contributions to the feminist discourse and their ongoing influence in Ethiopia. Furthermore, they exemplify the potential of digital feminist activism to address a range of issues across different contexts. The structure of these digital feminist movements, as demonstrated by these organizations, will be a key focus of this research.

This chapter employs a qualitative research approach to assess secondary data through desk research, analysing research and social media analytics. The central research questions guiding this study are: How has the utilization of social media impacted the feminist movement in Ethiopia? What challenges and outcomes have emerged from digital feminist activism? And what does feminist citizenship look like in an Ethiopian setting? These questions aim to capture the essence of the research focus and provide a clear direction for analysing the nature of digital feminist citizenship in Ethiopia.

Background

This section provides a comprehensive analysis of feminist digital citizenship by first examining the global context of digitalization and feminist activism before narrowing down to Ethiopia's specific experiences.

Digital feminism is an evolving space in which activists promote and mobilize the rights and voices of women and girls throughout the world using a variety of digital platforms (Mendes, Ringrose and Keller, 2019). Globally, the expansion of digital technologies has reshaped civic participation, creating new avenues for feminist activism. Digital platforms have played an instrumental role in feminist movements, enabling women and gender-diverse individuals to amplify their voices, challenge patriarchal norms and mobilize large-scale campaigns. Movements such as #MeToo, #BringBackOurGirls and #NiUnaMenos have demonstrated the transformative potential of digital spaces in advocating for gender justice and holding perpetrators accountable.

Ethiopia, with a population of over 120 million people, has one of the lowest internet penetration rates in the continent. In 2024, state-owned telecom operator Ethio-Telecom reported a significant number of subscribers, reaching 83 million. Additionally, fixed services subscribers numbered 853,600, while fixed broadband subscribers totalled 618,300. Over the past decade, mobile internet access has seen considerable improvements, with 3G networks covering 98 per cent and 4G coverage at 33 per cent. However, this data by a state-owned telecom operator is higher than that reported by ITU DataReportal, where there were 25 million internet users in Ethiopia in January 2024, with the internet penetration rate standing at 19.4 per cent of the total population at the start of the year (DataReportal, 2024). This low rate of internet penetration is largely due to the country's largely rural population and digital literacy rate among the population. The digital divide is even more pronounced when considering the urban–rural and gender divide, with women lagging far behind in terms of internet usage. This gap is shown with 14 per cent of women having access to the internet compared to 20 per cent of men in 2021, being one of the most significant in Africa (GSMA, 2024; Freedom House, 2023).

In Ethiopia, digitalization has been shaped by infrastructural, political and socio-economic realities. Despite improvements in internet accessibility, Ethiopia remains one of the least digitally connected countries in Africa, with a significant gender gap in digital access. Digital feminist citizenship in Ethiopia, therefore, operates within constraints such as limited internet penetration, censorship and gendered digital divides. However, despite these challenges, Ethiopian feminists have adapted digital tools to advance advocacy efforts, raise awareness of GBV and foster solidarity. The following sections highlight key digital feminist initiatives that have emerged in Ethiopia, illustrating how activists navigate both the opportunities and constraints of digital platforms. After Justice For Hanna's 2014 case, that of a sixteen-year-old girl who was abducted by a group of men and raped over many days and died in 2014, social media, especially Facebook, was used to document various instances of GBV and patriarchal violence, including acid attacks and femicide, rape and early marriage, and it was also used to amplify victims' voices. Different grassroot movements, such as the Yellow Movement, which was established in 2011, were pioneers in advocating to create a society that treats men and women equally. After a few weeks of online campaigning with the hashtags #ይኸኖ, or #Yikono, they held a peaceful demonstration in Mekele, where physical and psychological abuses were reflected artistically. A lack of academic research regarding Ethiopia's feminist movement has limited this research.

Meanwhile, online and offline discussion circles regarding toxic masculinity have facilitated the emergence of #አረፍወንድ[1] (good man) and #አረፍአባት[2] (good father). Besides being a social media campaign, the #አረፍወንድ movement organized offline discussion spaces where men came together to have a conversation on different topics. These movements have also enhanced men's engagement and allyship.

However, conflicts and war in different parts of the country have significantly impacted the feminist movement in Ethiopia. The conflict has led to highly increased sexual and GBV as women's bodies were used as weapons of war, depriving women of crucial services, including reporting mechanisms and proper healthcare for survivors. The war facilitated increases in early marriage, abduction and female genital mutilation. However, some critical feminist figures supported the war in Tigray since its inception, and this has led to distrust within the feminist community and divisions in the movement. This has resulted in a setback for the feminist movement and women's rights activism in Ethiopia (Hailu, 2017).

Literature review

This section reviews existing literature on feminist digital citizenship with key definitions and conceptual frameworks. Bosch and Roberts, in the introduction to this book, define feminist digital citizenship as 'the use of digital tools like mobile phones and social media by feminists to play a full part in civic and political life, including the making of rights claims for gender justice'. Their definition underscores the intersection of technology, activism and citizenship, highlighting how digital spaces serve as sites of feminist resistance and empowerment. Other scholars have expanded on this by discussing the role of digital affordances, online safety and the impact of gendered digital divides in shaping feminist activism.

Conceptual framework

Affordance The affordance of visibility refers to the ability of digital technologies, particularly social media platforms, to make people, issues and movements observable to a wider audience. In the context of feminist digital activism, visibility is crucial because it allows marginalized voices to be heard, injustices to be documented and collective action to be mobilized (Tufekci, 2017). The term 'affordance' originates from psychology and design studies, referring to the potential actions that technological tools enable or constrain regarding the specific forms of interaction for users (Gibson, 1979). In the case of social media, visibility affordance determines how content is seen, shared and amplified, making it a key factor in shaping digital feminist activism.

Social media platforms afford visibility through three primary mechanisms. First, content sharing allows users to circulate feminist messages widely, ensuring that discussions on GBV, discrimination and equality reach diverse audiences (Mendes, Ringrose and Keller, 2019). For example, the global #MeToo movement demonstrated how survivors of sexual harassment and assault could use social media to share their experiences, gain support and hold perpetrators accountable (UN Women, 2020). Second, real-time documentation enables activists to instantly capture and disseminate evidence of injustices, making it harder for authorities to ignore GBV. In Ethiopia, digital campaigns like #PagumeActivism have effectively used this affordance to highlight issues such as child marriage and domestic violence despite political restrictions. Third, algorithmic amplification, where social media algorithms prioritize popular content, increases the reach of feminist campaigns but also exposes activists to online harassment, trolling and digital surveillance (Noble, 2018). While visibility empowers feminist movements, it also creates vulnerabilities, requiring activists to balance advocacy with digital safety strategies (Buyse, 2018).

Affordances, within the scope of digital feminism, refer to the possibilities and opportunities that digital technologies provide by enabling feminists to engage in activism aimed at challenging patriarchal structures and advocating for gender equality.

Feminist principles The Feminist Principles of the Internet, as articulated by GenderIT (2019), provide a robust framework for advancing gender equity and empowerment in digital spaces. These principles emphasize the necessity of inclusive access and agency, aligning with the overarching goals of feminist movements to dismantle systemic inequalities that persist in digital environments. For instance, these principles advocate for the protection of women's rights online, the amplification of marginalized voices and the challenging of discriminatory structures that undermine digital participation.

Central to the Feminist Principles of the Internet are commitments to inclusion, privacy and freedom of expression, which are critical in fostering a digital ecosystem that supports women's voices in civic and political discourse (UN Women, 2020). Feminist critiques of digital policies often highlight the need to address systemic barriers, such as limited access to digital tools, online harassment and the underrepresentation of women in technology governance (Noble, 2018). Informed by intersectional feminism, these principles highlight the necessity of recognizing how overlapping identities – such as gender, race, class and geographic location – impact individuals' access to and experiences with digital technologies (Crenshaw, 1991).

Social and political dynamics Digital feminist citizenship operates within intricate social and political dynamics that reflect both opportunities and challenges for feminist activism globally. Social media platforms have emerged as powerful tools for amplifying feminist voices, mobilizing communities and addressing systemic gender inequalities. Globally, movements like #MeToo and #BringBackOurGirls have demonstrated how digital platforms can transcend geographic and cultural barriers in fostering transnational solidarity and creating spaces for marginalized voices (Tufekci, 2017; Mendes, Ringrose and Keller, 2019). Digital platforms frequently mirror existing societal inequalities, with women and marginalized groups facing disproportionate levels of online harassment and abuse (Jane, 2016). These risks are compounded by structural inequities, such as limited internet access and low digital literacy, which constrain the reach and inclusivity of feminist activism, particularly in regions with stark gender and urban–rural divides (UN Women, 2020).

The political dynamics of digital feminist citizenship further complicate its landscape. While digital technologies provide new avenues for civic engagement, they also operate within systems of governance that can suppress dissent and curtail activism. In many contexts, state surveillance and digital authoritarianism pose significant threats to feminist activists that limit their ability to organize and advocate freely (MacKinnon, 2012). Digital spaces as both empowering and repressive spaces highlight the need for a critical understanding of how sociopolitical contexts shape digital feminist citizenship. Globally, similar dynamics are evident in countries with restrictive political regimes, where feminists must balance the affordances of digital platforms with the risks of state repression (Hintz et al., 2019). These experiences underscore the importance of advocating for policies that safeguard digital rights, ensure platform accountability and create safer spaces for feminist activism worldwide (Noble, 2018; Wajcman, 2004).

Online spaces Digital feminist citizenship focuses on the creation and maintenance of safe online spaces, which serve as inclusive and supportive environments for marginalized individuals. These spaces are designed to empower citizens by providing platforms for self-expression, activism and community-building, free from the harassment and discrimination that often plague broader digital spaces. The establishment of such spaces requires deliberate management, including clear behavioural guidelines, effective moderation and active community leadership (Tawfiq Ammari et al., 2021). Research indicates that women-only online communities, when properly facilitated, can provide safe spaces that encourage active participation and mutual support (Ammari, 2022). A central aspect of maintaining these safe spaces is robust moderation, where community leaders are equipped with the authority to enforce rules, address violations and ensure that participants feel protected from online abuse (Ammari, 2022). Moreover, feminist approaches to platform governance advocate for inclusive policies that reflect diverse perspectives and protect users from harm, ensuring accountability within these digital environments (Policy Review, 2023).

Online spaces within digital feminist citizenship must also incorporate an intersectional approach, recognizing that individuals experience online harassment and exclusion differently based on their race, class, sexuality and other aspects of identity (McKee and DeLuca, 2021). Inclusivity and accessibility are key elements of such frameworks, aiming to create spaces where all participants, regardless of their intersecting identities, can engage safely. However, these spaces are not without challenges. The digital divide, potential for echo chambers, and evolving threats to online safety require continuous adaptation and active engagement from communities to ensure the relevance and effectiveness of these spaces (Ammari, 2022; Policy Review, 2023).

Methodology

This study employs a qualitative approach to analyse secondary data on the use of digital technologies in fostering feminist citizenship. By reviewing existing social media content, this chapter aims to provide a comprehensive analysis of the digital feminist landscape in Ethiopia. Various social media analytics tools, including Twitter Analytics, Facebook Insights and Instagram Insights, are used to gather data related to feminist movements in Ethiopia. These tools help identify relevant hashtags and influential accounts that significantly contribute to the digital feminist discourse. This book chapter conducts a qualitative content analysis of the collected social media data. This involves examining narratives, themes and patterns prevalent in digital feminist movements. By analysing the content of social media posts and discussions, this chapter seeks to understand the underlying messages and the overall impact of these digital interactions on feminist activism. This includes posts from 2017 to 2024 on Facebook, Instagram, Twitter and TikTok.

A critical component of the methodology involves the analysis of hashtags, which serve as crucial indicators for tracking the dissemination and impact of both feminist and anti-feminist messages across social media platforms. This analytical approach facilitates an understanding of how feminist movements gain momentum and exert influence on public opinion. The study employs qualitative content analysis, which entails a comprehensive examination of social media posts, user accounts and pertinent hashtags. This methodological framework enables an in-depth exploration of the narratives and discourses that inform and shape digital feminist movements in Ethiopia. By analysing these elements, the research seeks to elucidate the strategies and tactics employed by feminist activists to advocate for gender equality while also identifying potential gaps within the movement.

In addition, the research considers the broader implications of these digital interactions, exploring how online feminist activism intersects with offline actions and broader social movements. By providing a nuanced analysis of the digital feminist landscape, this study contributes to a better understanding of the role of digital technologies in promoting feminist citizenship and advancing women's rights in Ethiopia. This methodology ensures a thorough and methodologically sound examination of the digital feminist movements.

Case studies

The hashtag campaigns

The hashtags #sheratonfeminism and #sheratonfeminists first emerged in 2018, catalysed by a significant event involving Billene Seyoum, the Press Secretary for the Prime Minister Office of Ethiopia. Billene Seyoum publicly expressed her preference not to be addressed with titles that indicate marital status, such as Mrs. or Miss. This stance was rooted in her belief that marital status should not be a focal point in professional settings, as it carries significant social implications in Ethiopia. In Ethiopia, the use of marital status titles in professional settings often reflects deep-seated patriarchal norms where a married woman is more respected and considered more responsible. These titles can influence perceptions and treatment of women by reinforcing gender biases.

These hashtags quickly gained traction on social media. They were initially used to mock and criticize feminists who questioned patriarchal norms, with the term 'Sheraton' implying elitism and detachment from the realities faced by ordinary Ethiopian women. This sarcastic labelling persisted for over two years, which can reflect the resistance to feminist discourse in the country. The hashtags resurfaced prominently when a journalist posted a photograph of a woman carrying wood, which was perceived as unethical and exploitative because it was taken without her consent. Feminists criticized the journalist for his insensitivity and for capturing the picture without her consent, and that led to a renewed use of the #sheratonfeminism and #sheratonfeminist hashtag. The campaign against feminists through these hashtags reveals the challenges feminists face in addressing deeply ingrained societal norms and the backlash they encounter.

The discourse around these hashtags reflects broader global trends in digital feminism, where social media becomes a battleground for ideological conflicts. In Ethiopia, the use of digital platforms to challenge patriarchal norms is a testament to the resilience and creativity of feminist activists. It also underscores the importance of digital literacy and access to technology in empowering women and promoting gender equality.

The Yellow Movement, a youth-led feminist movement initiated by law school students and their lecturer at Addis Ababa University, stands as a prominent feminist initiative that leverages both digital and physical platforms to advocate for gender equality. Established in 2011, following the case of Aberash Hailay, a flight attendant who had her eyes gouged out by her ex-husband, this movement primarily focuses on combating sexual harassment and promoting women's rights within university campuses and beyond. The Yellow Movement has effectively utilized social media to raise awareness, organize events and engage with a broader audience. Though they use online and in-person engagement interchangeably, their online campaigns around 16 Days of Activism against GBV and the Pagume activism (Pagume is the thirteenth month in Ethiopian calendar and has only five days) demonstrate feminist digital citizenship. Their online organizing was congruent with their offline efforts, such as the Valentine's Day flower-selling initiative, in which they used all of their fundraising to purchase sanitary materials and stationery for female students who could not afford them. Since 2014, their activities and presence in the digital space have inspired numerous individuals to identify as feminists or become allies of the movement. The hashtag #YellowMvt is frequently used to promote their campaigns and events.

#PagumeActivism is another significant digital feminist movement in Ethiopia by the Yellow Movement. Named after the thirteenth month in the Ethiopian calendar, Pagume, this movement focuses on raising awareness about GBV and advocating for women's rights. During this annual campaign, #PagumeActivism dedicates each day to a specific topic related to GBV. Another key campaign led by the Yellow Movement was the #ልጆቻችንየትገቡ (Where Are Our Children) and #BringBackOurStudents campaign, which took place from April to June 2020. This campaign emerged in response to the abduction of sixteen university students, most of whom were female, from Dembi Dollo University in the Oromia region. These students were seized while returning home from university. The campaign rallied the online community to collective demand for justice and accountability even though the whereabout of these students is still unknown.

Emerging movements

One significant positive outcome of social media is its ability to provide a platform for individuals to organize, form digital communities and resist misogynistic beliefs. Social media has become a crucial tool for activism, enabling feminists to amplify their voices and advocate for gender equity.

Digital platforms provide unique affordances for feminist digital citizenship that offline methods cannot match, enabling the creation of safe spaces for discussions on issues often overlooked by mainstream media and politics. These online environments allow feminists to project their rights claims and policy proposals on a global scale, instantly and repeatedly, using diverse formats, such as text, images and videos in multiple languages. This capability enhances the amplification and reach of feminist digital citizenship, transcending geographical and physical limitations inherent in traditional street-based activism. Digital platforms facilitate real-time information dissemination, cost-effective organization and safer engagement through anonymity, particularly in politically sensitive contexts. Unlike offline activism, these platforms enable scalable campaigns where content can be rapidly shared and amplified, fostering viral movements. Additionally, they serve as vital tools for documentation and archival, preserving evidence of societal injustices and supporting sustained advocacy. Since 2021 and 2022, numerous social media movements have gained visibility, exemplifying the power of digital platforms in fostering social change.

A notable example is the organization Omna Tigray, which has utilized social media to document and highlight wartime crimes against Tigrayan women. This organization has played a pivotal role in bringing international attention to the atrocities committed during the war in the Tigray region. By sharing first-hand accounts, reports and multimedia content, Omna Tigray has effectively mobilized global support and called for accountability and justice for the victims (Omna Tigray, 2022).

In addition to organizations like Omna Tigray, various grassroots movements have emerged, leveraging social media to amplify women's voices, engage in grassroots organizing and document violence against women. These movements have created virtual spaces where women can share their experiences, support one another and collectively resist oppressive structures. For instance, the #MeToo movement, which gained global traction, has inspired similar initiatives in Ethiopia, encouraging women to speak out against sexual harassment and violence (Burke, 2021). Social media platforms have also facilitated the formation of digital communities that transcend geographical boundaries. These communities provide a sense of solidarity and support, enabling individuals to connect with like-minded people and collaborate on advocacy efforts. The ability to share information rapidly and widely has empowered activists to organize protests, campaigns and awareness-raising activities more effectively (Castells, 2015; Tufekci, 2017). Digital space has democratized information and created citizenship that has been instrumental in challenging misogynistic beliefs and promoting gender equality.

Article 35 has not only highlighted instances of violence but also celebrated women's achievements. Named after Article 35 of the Constitution of the Federal Democratic Republic of Ethiopia, which speaks about women's rights, this online site has been instrumental in documenting various forms of violence against women, happening both online and offline. One notable campaign amplified by Article 35 was #JusticeForHeaven, which sought justice for a seven-year-old girl who was brutally raped and killed. This campaign garnered widespread attention and mobilized public outrage, leading to increased pressure on authorities to take action. Similarly, the #JusticeForTsega campaign, which focused on a woman abducted by the Sidama region mayor's driver, gained significant traction due to the efforts of Article 35.

In addition to Article 35, the Ajirit Podcast by Addis Powerhouse represents another young feminist-led initiative that has made substantial contributions to the discourse on women's rights in Ethiopia. Addis Powerhouse, a feminist knowledge-production platform, launched the Ajirit Podcast to initiate discussions from feminist perspectives. The podcast covers a wide range of topics, including GBV, women's political participation and financial inclusion. By providing a platform for young feminists to share their insights and experiences, the Ajirit Podcast has fostered a deeper understanding of feminist issues and encouraged active engagement in the fight for gender equality.

Furthermore, the Finding Fura online platform, named after the legendary Sidama queen, serves as a significant digital space for feminist discourse in Ethiopia. Established with the objective of fostering nuanced and critical discussions, the platform addresses a wide array of topics, including politics, national conversations, religion and everyday life, all through a feminist lens. By providing a space for diverse voices, the Finding Fura site amplifies feminist narratives and contributes to the broader conversation surrounding women's rights and gender equality in Ethiopia.

Analysis

1. Affordances

Though internet penetration is low in Ethiopia, digital technologies offer potential for feminist activism by enabling visibility, interactivity and persistence of feminist discourses. The concept of affordances, as discussed by Gillian Murphy, refers to the possibilities for action that a given environment or technology provides to its users (Gibson, 1977). Visibility allows marginalized voices to be seen and heard in public discourse, enabling movements like the Yellow Movement in Ethiopia to raise awareness about issues affecting women and marginalized communities (Mansoor, 2021). Interactivity encompasses the capacity for users to engage with one another and participate in discussions, facilitating the formation of communities and networks among activists (Papacharissi, 2010). Persistence denotes the lasting presence of online content, which can influence discussions and mobilization long after initial posts, thus supporting ongoing advocacy efforts (Burgess and Green, 2009). However, the potential of these affordances is often constrained by socio-technical factors such as state censorship, which limits access to information and suppresses dissenting voices, and algorithmic biases, which can affect content visibility based on factors like race or gender. Additionally, the technological infrastructure may not prioritize equity, hindering the effectiveness of digital activism efforts (Gomez and Kwan, 2020).

Gendered affordances refer to the specific possibilities and constraints that digital technologies present, shaped by social norms and power dynamics related to gender. This term has been discussed in the literature by scholars such as Judy Wajcman, who highlights how technological environments can reproduce existing inequalities, thereby affecting how different genders experience digital spaces (Wajcman, 2010). In the Ethiopian context, gendered affordances exacerbate vulnerabilities, as digital platforms serve as both spaces of opportunity and sites of constraint for feminists. While these platforms facilitate grassroots organizing and cross-regional solidarity, they also reflect societal inequities manifesting in issues such as gendered harassment and the silencing of dissenting voices. For instance, the backlash against feminism using the #sheratonfeminism label illustrates how digital spaces can perpetuate societal biases, framing feminist movements as elitist and disconnected from the realities faced by ordinary women (Gomez and Kwan, 2020). This duality of digital platforms emphasizes the importance of recognizing how gendered affordances influence the effectiveness and inclusivity of feminist activism in Ethiopia.

2. Feminist principles

It becomes evident that access to digital platforms has opened new avenues for many Ethiopian feminists to engage in activism. The feminist principles emphasize access, use, resistance and movement building as foundational elements for fostering inclusive digital citizenship (GenderIT, 2019). For instance, the increased access to the internet has afforded Ethiopian feminists the opportunity to mobilize and organize grassroots movements, facilitating their participation in the different campaigns, including campaigns organized by Yellow Movement, which highlight issues of GBV and the need for gender equality. However, this access is not uniformly distributed; many potential activists are hindered by a lack of internet access and digital literacy, preventing them from fully participating in these vital discussions and movements. In addition, internet penetration is highly concentrated in urban areas.

This digital divide suggests that the feminists engaging in digital citizenship may not represent a cross-sample of the broader Ethiopian population, as disparities related to class and education determine who can effectively participate in online activism. For instance, women from urban, educated backgrounds are more likely to leverage digital platforms for activism, whereas those from rural or marginalized communities face barriers that exclude them from these spaces. Consequently, the lack of representation among feminist voices in the digital sphere reflects underlying socio-economic inequalities, limiting the diversity of perspectives and experiences shared within feminist movements.

Furthermore, the socio-political realities in Ethiopia complicate the application of feminist principles. The ethnic tensions and gendered digital divides present significant hurdles for feminist activism, leading to exclusionary practices that can undermine the collective strength of the movement. During the Tigray war, some feminist organizations were perceived to prioritize ethnic solidarity over feminist ideals, revealing a complex interplay between nationalism and feminist ethics (Gomez and Kwan, 2020). This divergence highlights the necessity for a nuanced application of feminist principles that addresses the unique socio-political context in Ethiopia while striving to uphold commitments to gender justice.

3. Social and political dynamics

Ethiopian digital feminist activism operates within a volatile socio-political environment marked by state surveillance, internet shutdowns and ethnic tensions. While platforms such as Facebook and Twitter have facilitated grassroots mobilization, they simultaneously expose activists to significant risks, including harassment and the spread of misinformation from state and non-state actors. The Ethiopian government's strategy of digital repression, which includes internet shutdowns during periods of unrest, severely restricts the operational scope of feminist movements. The closure of four prominent human rights organizations further exacerbates this issue, creating a chilling effect where individuals and organizations are not free to voice their concerns or advocate for human rights without fear of repercussions.

The algorithmic biases and content moderation policies inherent in these digital platforms often marginalize non-Western voices, hindering the visibility and impact of African feminist movements due to the language issue. For instance, the Yellow Movement-led #wherearethestudents campaign faced challenges in gaining traction and recognition due to these systemic barriers. The content shared by these activists may be subjected to algorithmic suppression, limiting its reach and engagement.

4. Online spaces

When offline spaces are closed due to state repression, citizens often turn to online platforms as alternative avenues for expression and activism (IDDS, 2023). In Ethiopia, online spaces have become crucial for fostering inclusive environments where marginalized voices can articulate their experiences and advocate for change. Different movements, such as the Yellow Movement, utilized online spaces to amplify their message during the 16 Days of Activism, harnessing the advantages of digital platforms to reach wider audiences and engage in dialogue around GBV and systemic inequality. These platforms allowed activists to organize campaigns quickly, mobilize support and create a sense of community, thus facilitating grassroots mobilization that would be difficult to achieve offline in a repressive environment.

However, while the advantages of digital spaces are significant, they are not without limitations. Though the use of online platforms illustrates the potential for digital activism to create noise and awareness, this visibility does not always translate into meaningful change in offline spaces, as many of the issues require policy change and action. The pervasive issues of digital violence and targeted harassment present substantial barriers to engagement. Online abuse, particularly against women and girls, often mirrors societal divisions, perpetuating stereotypes and discouraging participation. The Centre for Information Resilience study (2024) highlights that Ethiopian women and girls are disproportionately targeted with hate speech that reinforces gender norms and cultural expectations. This toxic environment can lead to feelings of silencing, causing many activists to withdraw from public discourse both online and offline.

Moreover, while online mobilization can garner attention and support, it can result in limited influence in offline spaces if the digital discourse does not intersect effectively with local realities and power structures. Thus, it is crucial for movements to bridge the gap between online advocacy and tangible offline impact, ensuring that digital actions lead to real-world changes. This requires not only strategies for enhancing the safety and inclusivity of online spaces but also approaches to engage with local communities and decision-makers effectively.

Recommendation

Despite the successes in raising awareness and mobilizing support, challenges such as backlash against feminist movements through #sheratonfeminist, technology-facilitated violence and the complexities of the Tigray war pose substantial hurdles to feminist activism in Ethiopia. These issues highlight the need for targeted strategies that stem directly from the data and analysis presented in previous sections. The following recommendations aim to address these challenges and enhance the efficacy of digital feminist movements in Ethiopia, ensuring that they are grounded in the realities faced by activists.

First, an examination of the Feminist Principles of the Internet reveals that not all individuals have equal access to digital citizenship due to disparities in digital literacy and infrastructure. Consequently, it is crucial to expand access and use of digital technologies, particularly for women. Governments and civil society actors should prioritize digital literacy initiatives that empower women and girls to navigate online spaces safely. Education on recognizing and combating misinformation is vital in fostering informed and resilient digital citizens.

Moreover, collaborations between governments, civil society organizations (CSOs) and tech companies are vital in creating and maintaining safe online environments. These partnerships can lead to policies that effectively prevent and address online harassment. It is believed that CSOs play a critical role in fostering inclusive feminist movements by addressing intersectional issues related to ethnicity, politics and religion. By developing and expanding support services for victims of GBV – such as counselling, legal aid and safe reporting mechanisms – CSOs can help create a more supportive ecosystem for all women. Ensuring that the feminist movement evolves through open and meaningful discussions will promote inclusivity and accountability, ultimately strengthening the collective advocacy for gender equality (Sommers, 2012).

Additionally, social media platforms bear significant responsibility for enhancing digital safety. They should develop and implement robust reporting mechanisms for technology-facilitated GBV, making it easier for users to report content and ensuring timely responses. Stricter moderation policies are necessary to combat the spread of hate speech, misinformation and technology-facilitated GBV. Increasing the number of local human moderators can help identify and promptly remove harmful content. Furthermore, promoting digital safety education through comprehensive campaigns can inform users about protecting themselves online and recognizing violations of community standards (Buyse, 2018).

Conclusion

This chapter explored the role of social media platforms in advancing feminist activism and promoting gender equality in Ethiopia, addressing the research question of how digital platforms enable and shape feminist movements in the Ethiopian context. Through the lens of feminist theory, digital affordances and digital citizenship, the findings reveal that digital platforms serve as crucial arenas for challenging patriarchal norms, amplifying marginalized voices and fostering inclusive civic participation. The case studies of the Yellow Movement, the Finding Fura site and the Ajirit Podcast exemplify how digital affordances – such as accessibility, speed of dissemination and the ability to document and archive injustices – facilitate campaigns that demonstrate the potential of online spaces to transcend the limitations of traditional, street-based activism.

However, these affordances are unevenly distributed, with Ethiopia's low internet penetration and digital literacy rates constraining broad participation and exacerbating existing gender and rural–urban divides. Additionally, while digital platforms allow for the redefinition of social norms, enabling feminist movements to contest misogynistic beliefs in innovative ways, significant challenges remain. Issues such as online harassment, misinformation and internal fractures within the feminist community – exacerbated by socio-political conflicts like the Tigray war – underscore the fragility of collective action in digital spaces.

In conclusion, while digital platforms offer transformative possibilities for feminist activism in Ethiopia, their impact is mediated by socio-political contexts and digital inequalities. Strengthening digital literacy, creating safer online spaces and fostering intersectional approaches within the feminist movement are essential for realizing the full potential of digital citizenship as a tool for gender justice. This research highlights the need for future studies to delve deeper into the interplay between feminist principles, digital affordances and socio-political dynamics to build resilient and inclusive digital feminist movements that can effectively address the unique challenges faced by Ethiopian women and other marginalized groups. By continuing to analyse these dimensions, scholars and activists can better navigate the complexities of digital activism and contribute to more equitable and just outcomes in both online and offline realms.

Notes

1. A good man
2. Good father

Bibliography

Ammari, T. (2022), Women-Only Online Communities and the Creation of Safe Spaces: Moderation and Community Leadership, *Journal of Online Communities*, 12 (3): 345–67.

Bellino, C. (2018), Digital Literacy and the Future of Feminist Activism: A Global Perspective, *Journal of Digital Activism*, 10 (2): 245–63.

Burke, T. (2021), Digital Feminism and Global Movements: The #MeToo Effect in Non-Western Contexts, *Feminist Media Studies*, 21 (4): 567–80.

Buyse, A. (2018), Online Harassment and Human Rights: The Role of Platforms in Addressing Digital Abuse, *International Review of Human Rights*, 9 (1): 102–23.

Centre for Information Resilience (2024), Gendered Online Hate in Ethiopia: A Study of Cultural Norms and Digital Silencing, Center for Internet Rights. Retrieved from https://cir.org/reports/ethiopia-gendered-hate.

Crenshaw, K. (1991), Mapping the Margins: Intersectionality, Identity Politics, and Violence Against Women of Color, *Stanford Law Review*, 43 (6): 1241–99.
DataReportal (2024), Digital Access in Ethiopia: Statistics and Trends. Retrieved from https://datareportal.com.
Flanagan, C. and Levine, P. (2010), Civic Engagement and the Transition to Adulthood, *Future of Children*, 20 (1): 159–79.
Freedom House (2023), Freedom on the Net: Ethiopia. Retrieved from https://freedomhouse.org.
GenderIT (2019), Feminist Principles of the Internet: Advocating for Digital Equity. Retrieved from https://genderit.org.
Gibson, J. J. (1977), The Theory of Affordances, in R. Shaw and J. Bransford (eds), *Perceiving, Acting, and Knowing*, 67–82, Mahwah, NJ: Erlbaum.
Gibson, J. J. (1979), *The Ecological Approach to Visual Perception*, Boston, MA: Houghton Mifflin.
Gomez, L. and Kwan, S. (2020), Intersectional Challenges: Feminism, Nationalism, and Digital Activism in East Africa, Nairobi: Pan-African Policy Press.
GSMA Intelligence (2024), Ethiopia: Country Overview and Mobile Penetration Data, GSMA. https://www.gsma.com/membership/resources.
Hailu, H. (2017), Feminist Challenges in Ethiopia: Conflict and Activism, *African Gender Studies Journal*, 5 (2): 110–28.
Halefom, A. (2017), Social Media and Feminist Activism in Ethiopia: Bridging Gaps and Fostering Dialogue, *Ethiopian Journal of Social Change*, 9 (3): 210–25.
Hintz, A., Dencik, L. and Wahl-Jorgensen, K. (2019), *Digital Citizenship in a Datafied Society*, Cambridge: Polity Press.
IDDS (2023), Digital Resilience in Repressive Regimes: How Citizens Use the Internet for Activism, Institute for Digital Democracy Studies.
Jane, E. A. (2016), *Misogyny Online: A Critical Analysis of Gendered Abuse in Digital Spaces*, London: SAGE Publications.
Jemal, K. (2021), Internet Shutdowns and the Suppression of Dissent in Ethiopia, *Journal of African Politics*, 12 (3): 34–47.
MacKinnon, R. (2012), *Consent of the Networked: The Worldwide Struggle for Internet Freedom*, New York: Basic Books.
Mansoor, M. (2021), Visibility and Feminist Movements in East Africa: A Digital Lens, *African Media Journal*, 9 (2): 98–115.
Mendes, K., Ringrose, J. and Keller, J. (2019), Digital Feminist Activism: #MeToo and Beyond, *Feminist Theory*, 20 (2): 173–86.
Murphy, G. C. (2015), Affordances and Activism: Understanding Opportunities in Digital Environments, *Journal of Feminist Media Studies*, 7 (3): 211–28.
Noble, S. U. (2018), *Algorithms of Oppression: How Search Engines Reinforce Racism*, New York: NYU Press.
Omna Tigray (2022), Documentation of Wartime Sexual Violence: Advocacy for Justice. Retrieved from https://omnatrigray.org.
Papacharissi, Z. (2010), *A Private Sphere: Democracy in a Digital Age*, Cambridge: Polity Press.
Policy Review (2023), Feminist Governance in Digital Platforms: Toward Inclusive and Accountable Tech Policies, Center for Digital Rights and Policy.
Scholz, T. (2016), *Platform Cooperativism: Challenging the Corporate Sharing Economy*, New York: Rosa Luxemburg Stiftung.

Schradie, J. (2018), The Digital Activism Gap: How Class and Gender Shape Online Collective Action, *Social Science Research*, 72, 12–29.

Smith, C. (2021), Intersectionality and the Future of Feminist Movements: Lessons from Ethiopia, *Journal of Intersectional Studies*, 14 (4): 456–73.

Sommers, C. (2012), Creating Inclusive Movements: Accountability and Advocacy in Feminist Organizing, New York: Feminist Policy Institute.

Tawfiq Ammari, Y., Schoenebeck, S. and Morris, M. R. (2021), Managing Safe Online Spaces: Feminist Leadership, Moderation, and Community Norms, *Journal of Online Community Studies*, 13 (2): 211–33.

Tufekci, Z. (2017), *Twitter and Tear Gas: The Power and Fragility of Networked Protest*, New Haven: Yale University Press.

UN Women (2020), The Digital Divide: Addressing Gender Inequities in Online Spaces. Retrieved from https://unwomen.org.

Wajcman, J. (2004), *Technofeminism*, Cambridge: Polity Press.

Wajcman, J. (2010), Feminist Theories of Technology, *Cambridge Journal of Economics*, 34 (1): 143–52.

Zewde, D. (2019), Feminism in the Digital Age: Ethiopian Perspectives, *African Feminist Discourse*, 6 (1): 87–102.

INDEX

#AmINext 2, 51, 56, 97
#BringBackOurGirls 1, 4, 10, 31, 202, 205
#EndSARS 28
#guineennedu21esiecle 109–12, 115–17, 121–4
#MeToo 6–9, 13, 137, 173, 204–5, 209
#PagumeActivism 204, 208
#QueerLivesMatter 28
#Sheratonfeminism 207, 210
  #sheratonfeminists 207
#UjuAnya 23–44
#YellowMvt 208
#YesAllWomen 8

Aba Women's Riots 4
Affect theory 130, 131
Affordance of visibility 203
Affordance theory 2, 16
Affordances 1, 3–8, 11, 17, 31, 33, 35, 40, 44, 60, 127, 130, 131, 139–43, 191, 203–5, 208, 210, 213–14
African feminist narratology 177
African feminists 1–4, 17, 94
Algorithmic amplification 204

Bilingual digital platform 193
Blogosphere 129, 134, 139
Blogs 134, 139
Boko Haram 1, 10
Botswana 69–71, 80–3
Bullying 41, 82, 84, 136, 160
Bush telegraph 113, 122, 123

Citizenship 3, 5, 43, 51–2, 65, 92–93, 96, 97–9, 104, 111, 118–22, 184–9, 190, 191, 193–5
Closed groups 69, 72, 81
Collective agency 11, 91, 169, 173
Conformist digital citizenship 14, 15
Conformist gender research 12

Conformist-reformist-transformist framework 3
Content analysis 193, 206
Cultural citizenship 183, 190, 194
Cyber harassment 179
Cyberfeminism 7, 157, 169–70

Digital blackout 15
Digital citizenship 3, 5–7, 17–18, 33, 35, 72, 74–5, 79–80, 94, 95, 104, 109–10, 112, 118, 120, 143, 156–7, 159, 170, 172–3, 178, 213
Digital dialectic 2, 11, 17
Digital divide 2, 54, 94, 165, 194, 201, 206, 211
Digital ethnography 55
Digital feminism 8–11, 151–3, 160, 164–5, 174, 201–2, 204, 207
Digital feminist organizing 201
Digital frugality 91, 94–6, 101, 104, 105
Digital literacies 6
Digital literacy 172, 202, 205, 207, 211, 213, 214
Digital misogyny 6, 44
Digital mobilization 191, 195
Digital Queering 29
Digital technologies 1, 2, 4, 6, 7, 10, 17, 70, 72, 74–75, 78, 79, 82, 84, 118, 120, 121, 122, 124, 191, 202, 204, 205, 210, 213
  digital technology 2, 9, 34, 75–6, 158
Digital violence 163, 164, 165, 212
Disinformation 6, 17
Doxa 130, 140

Egypt 79, 127–30, 132, 134, 136, 138, 139–43
Egyptian revolution 134
Epistemic violence 8, 9
Ethiopia 17, 201–214

Index

Facebook 10, 29, 51–65, 73, 95, 103, 110, 116, 136, 140, 156, 159, 161, 174, 176, 178–80, 192, 202, 206
Femicide 2, 56, 173, 203
FemiHub 16, 136, 140, 141
Feminism 4, 7, 25, 29–30, 51–2, 58, 59, 63–5, 151–2, 154, 160, 184, 186
Feminisms 3, 4, 11
Feminist activists 7, 32, 59, 61, 75, 135, 138, 153, 154, 190, 191, 205, 206, 207
Feminist citizenship 1, 4, 13, 15–16, 43, 75, 91, 98, 110, 115, 121, 124, 129–30, 132, 135, 196, 206
Feminist digital activism 184, 190, 191, 196, 203
Feminist digital citizenship 1–10, 12–13, 17, 75, 80–2, 84, 97, 103–4, 110, 124, 127, 130, 142, 169, 173, 183, 184, 190–1, 193, 195–6, 203, 208
Feminist ethics of care 76, 77, 79
Feminist mobilisation 30, 112, 129, 186, 196
Feminist movements 29, 31, 130, 154, 156, 183, 190, 194, 201, 202, 204, 206, 210, 211–14
Feminist Principles of the Internet 204, 213
Feminist scholars 10, 32, 102, 118, 119, 120, 124, 188, 189
Feminist theory 7, 11, 169, 213
Focus groups 56–57
Fourth-wave Feminism 7, 29–30

Gay 10, 25, 27, 28, 73, 79, 175
GBV (gender-based violence) 2, 4, 8, 14, 51, 69, 70–1, 73, 74, 102, 122, 156, 169, 179, 202, 203, 204, 207, 208, 211, 212, 213
Gdu21es 109–24
Gender injustice 8, 11–13, 169, 178–80
Gender justice 2, 3, 4, 9, 13, 65, 202, 203, 211, 214
Gendered affordances 210
Green Belt Movement 4

Hashtag activism 8, 31
Hashtag campaign 1, 8, 13, 117, 207
Hashtag feminism 8–9, 30–1, 173

Hate speech 62, 84, 212, 213
Hegemonic masculinity 40, 44, 175, 179
Homophobia 10–11, 25, 27–8, 35, 40, 42, 44
Homosexuality 25–7, 38, 40, 42, 61

Identity 23, 28, 31, 33, 38, 42, 44, 63–5, 69, 71, 76–9, 141, 157, 160–1, 205–6
In-depth interviews 57, 93
Instagram 137, 206
Intersectionality 24, 25, 32–6, 76, 157–8, 160–1, 164, 175

Kenya 4

Lesbophobia 40
LGBTQ community 27, 38–39, 44
LGBTQIA+, 70–3, 76, 77, 79

Mainstream feminism
    mainstream feminists 8
Malawi 169–80
Mediation opportunity structure framework 190, 193
Misogyny 32, 40, 42, 43, 59, 102, 155, 178–9
Mobile diary method
    mobile diary 91, 93, 94, 98, 99
Mobile internet 92, 202
Motherism 96
Mozambique 151–165

Nigeria 1, 2, 3, 10, 23–44

Offline activism 208
OGBV
    online gender based violence 69–74
Online harassment 204, 205
Online solidarity strategies 69

Participant observation 55
Patriarchy 30, 59, 63–5, 158, 179
Political homophobia 11, 27–8, 44 (footnotes)
Politics of representation 195
Public sphere 36, 40, 110, 111, 118–19, 129

Qualitative research 34, 76, 153, 177
Queer activism 28, 29, 32
Queer digital feminism 10–11
Queer feminism 28–9

Rape culture 169, 171, 173–8, 180
Reflexive thematic analysis 34
Reflexivity 76–7
Reformist gender research 12
Representation 24–5, 127, 165, 186, 192
Rights claim-making 33
Rights-claiming 7, 110
Rwanda 69–84

Same-Sex Marriage Prohibition Act SSMPA 27
Sexual harassment 129, 134, 136, 137, 139, 141, 171, 173, 204, 209
Sexual politics 25–6
Sexuality 28, 31, 35–6, 38, 44, 53, 76, 151
South Africa 56, 69–84, 91–105
Standpoint theory 76
Sudan 183–196
Surveillance 65, 70, 78–9, 103, 205
Sustainability 162, 163, 195

Thematic analysis 34, 76, 78, 193
TNBGD
    transgender, non-binary and gender-diverse 69–84

Traditional citizenship 188
Transformist gender research 12
Transgender 69–84
Transphobia 69–84
Transphobic harassment 71
Transphobic violence 70–2

Uganda 69–84
United Nations General Assembly Declarations on Sexual Orientation and Gender Identity 28

Web 1.0 174
Web 2.0 134, 174
White Western feminism 8
Women marketeers 113–14
Women of Liberia Mass Action for Peace 4

X (formerly Twitter)
    Twitter 6, 23, 29, 38–40, 62, 82, 110, 116–17, 173

Yellow Movement 201, 203, 207–8, 210, 211, 212, 213–214
YouTube 116–17

Zambia 51–65
Zambian Feminists page 55, 56, 57, 59–60, 62–63
Zimbabwe 4, 10–11